FAMOUS ROBOTS & CYBORGS

FAMOUS ROBOTS & CYBORGS

An Encyclopedia of Robots from TV, Film, Literature, Comics, Toys, and More

DAN ROBERTS

Skyhorse Publishing

Copyright © 2014 by Dan Roberts

First published in Great Britain in 2013 by Remember When, an imprint of Pen & Sword Books Ltd

All Rights Reserved. No part of this book may be reproduced in any manner without the express written consent of the publisher, except in the case of brief excerpts in critical reviews or articles. All inquiries should be addressed to Skyhorse Publishing, 307 West 36th Street, 11th Floor, New York, NY 10018.

Skyhorse Publishing books may be purchased in bulk at special discounts for sales promotion, corporate gifts, fund-raising, or educational purposes. Special editions can also be created to specifications. For details, contact the Special Sales Department, Skyhorse Publishing, 307 West 36th Street, 11th Floor, New York, NY 10018 or info@skyhorsepublishing.com.

Skyhorse® and Skyhorse Publishing® are registered trademarks of Skyhorse Publishing, Inc.®, a Delaware corporation.

Visit our website at www.skyhorsepublishing.com.

10 9 8 7 6 5 4 3 2 1

Library of Congress Cataloging-in-Publication Data is available on file.

ISBN: 978-1-62636-389-2

Printed in China

Contents

Chapter 1:	Frequently Asked Questions	1
Chapter 2:	Milestones	5
Chapter 3:	The First Ones	15
Chapter 4:	The Best of the Worst	30
Chapter 5:	The Age of Paranoia	36
Chapter 6:	Double Trouble	54
Chapter 7:	A Golden Age	58
Chapter 8:	How To Kill A Robot	76
Chapter 9:	Sins And Revelations	78
Chapter 10:	We ♥ Robots	100
Chapter 11:	Transformations	102
Chapter 12:	Living Ships	124
Chapter 13:	The Modern Age	127
Chapter 14:	Danger, Humans!	139
Chapter 15:	Loving The Robot	141
Chapter 16:	The Perfect Robot	145

Chapter 1

Frequently Asked Questions

What is a robot?
An automaton or computer often in a vaguely anthropomorphic form, i.e. resembling the shape of a human. (Writer Isaac Asimov said 'It will be easier to be friends with human-shaped robots than with specialized machines of unrecognizable shape.') A robot is built to perform tasks which human beings find too boring or too difficult, or for which we simply do not have the physical or mental capacity.

What is a cyborg?
A 'cybernetic organism', partly robot and partly organic. It can be a humanoid who uses technology to a greater or lesser extent to enhance his/her physical existence by incorporating it with his/her human form, or a creature whose robotic circuitry is integrated with that of another organic being, e.g. an animal.

Why do so many stories about robots exist?
The automaton as a character is an endless source of inspiration for the storyteller in many media. Although the word 'Robot' itself can be traced back to Karel Čapek's 1920 play *R.U.R* (*Rossum's Universal Robots*), writers' fascination with automata can be seen in earlier works, for example by E.T.A Hoffmann and Villiers de l'Isle Adam. Humanity is interested in achieving the seemingly unachievable – can a robot be given the power by humanity to do things we cannot? The answer, of course, is yes, but how far will these abilities stretch….?

Could we one day invent a robot or cyborg which will replace us?
Brian Aldiss's short story *Who Can Replace A Man?* envisages this point, where intelligent machines have taken over the world. And what if the machines, our own creations, turned on us and we needed to fight back? It's a notion frighteningly brought into the twenty-first century by the *Terminator* film franchise.

Why do we find the robot so compelling?
Because they are like us, and yet not. They embody many aspects of the human form and yet are so distant in many ways from the experience of the human

condition. On one level, it must be that odd fascination with the almost-human, of what could be if our emotions and our spontaneity were somehow limited, controlled. The tensions between humanity and robots, that odd closeness-yet-distance which makes them like us yet so unlike us, has inspired many creative minds in film and television. The quirkiness of giving an automaton a close-to-human personality is a richly fertile area for amusement, whether the creation in question is tetchy and imperious like the Doctor's K-9, fussy and concerned like C-3PO, or simply quietly learning like *Star Trek*'s Data. Many of the robots we see here are compelling because they are so like, yet unlike, humanity – mimicking the form while possessing none of the substance. If only Humanity had rules; laws like those for robots as set down by Isaac Asimov…

What are Asimov's Laws?
1. Robots must never harm human beings or, through inaction, allow a human being to come to harm.
2. Robots must follow instructions from humans, but without violating rule 1.
3. Robots must protect themselves without violating the other rules.

Even as we read them today, we can see the potential problems and contradictions which might arise. What if a robot needed to protect itself, but this process involved attacking and destroying a hostile human being? On such dilemmas are the best sci-fi ideas built.

What can robots do that we cannot?
Data recall is a key robot trait. An enormous array of knowledge and memory is exhibited by automata both in fact and in fiction. Knowledge is power – knowing not just what human beings have done in the past but what they are doing now (CCTV, for example, presaged in such 'living computer' classics as *Colossus* and *Demon Seed*), and predicting where they might end up in any number of possible futures.

What can we do that robots cannot?
Feel – love, hate, fear… Emotion is what makes us human; the genuine capacity to allow our heart sometimes to rule our head. We have our animal instincts, but tempered with reason. Shakespeare's Iago expressed the balance thus: 'but we have reason to cool our raging motions, our carnal stings, our unbitted lusts.' (*Othello*, Act I, Scene 3.) How could a robot ever reproduce that balance? The Tin Man sought a heart because that, for him, represented humanity. *Star Trek*'s Data needs a special chip to provide him with artificial emotion. The Doctor in *Doctor Who* berates the Cybermen because they cannot remember when they last 'had the pleasure of smelling a flower, watching a sunset, eating a well-prepared meal.' Perhaps this

enables us to feel safe because no matter how advanced artificial intelligence becomes, it can never be truly human. The fulfilment of Alan Turing's test, a machine which cannot be distinguished from a human being, may be a long way off. Biologist Richard Dawkins, though, in *The Selfish Gene*, sees more in common between humans and robots in one sense: 'We are survival machines,' he says, 'robot vehicles blindly programmed to preserve the selfish molecules known as genes. This is a truth which still fills me with astonishment.'

What is an 'autonomous' robot?

One which functions under its own control instead of following programmed instructions or being controlled by a human operator. It reacts to its own environment and makes 'decisions' based on what it knows and has learnt. Additions such as vision, touch, and light sensors can help to work towards making a robot autonomous. To process information from a sensor, a robot 'brain' – microcontroller – is typically used.

Are robots frightening?

A good, efficient robot simply gets on with its task in an implacable manner, incapable of being flustered or of suffering the effects of stress, moods or fatigue. It should be there in the background, beavering away, quietly sparing Humanity from the tasks which are too safe, too boring, too repetitive. But the very anonymity of robots is what can cause them to be a little unsettling to us. Writers throughout the twentieth century have played on the tension between the anthropomorphic 'look' of a robot, its lack of the body language signals we read in humans, and the terrifying blankness that exists where we would expect to read doubt, fear, love, hate. This is automatonophobia – the fear of robots and, indeed, of anything else which superficially looks like a human being but does not behave like one, e.g. a ventriloquist's dummy, a waxwork, etc. We cannot cajole a robot, plead with it, persuade, bribe, or threaten it. The robot simply does what it is logically programmed to do. The cyborg is more complex, existing where flesh meets metal, where synapses meet circuitry. It has, ideally, the best qualities of both humanity and technology – or perhaps the worst. What happens at the interface of the body and robotics? How do the two work together, and where does the 'irrational' organic take over from the cold, calculating computer? Again, this has been fertile ground for sci-fi writers over the years.

What do the experts say?

'The best part of robotics is getting something to work using the efforts of many different people and skills. In robotics, there's a lot of teamwork. No one person does all the work to make a robot operate. It's always interesting to me to see all the different skills required to make our robots work.' – NASA Robotics Engineer Linda Kobayashi.

4 Famous Robots & Cyborgs

'Here is something to keep in mind. When Henry Ford introduced the Model T, there were essentially no cars on the road. Ten years later, his company alone was producing two million cars a year. And that was a century ago… Once the fundamentals of robotic intelligence are figured out – vision being the key limiter right now – robots will spread with startling speed. They will be a commodity, just like cellphones or desktop computers.' – Author Marshall Brain, founder of the website *How Stuff Works*.

It's worth adding here that IBM's chief executive, Thomas J. Watson, is *alleged* to have said in 1943, 'I think there is a world market for maybe five computers.' Who knows how far and how fast robot culture will spread, or by how much we have underestimated its potential reach?

'Machines smart enough to do anything for us will probably also be able to do anything with us: go to dinner, own property, compete for sexual partners. They might even have passionate opinions about politics or, like the robots on Battlestar Galactica, even religious beliefs. Some have worried about robot rebellions, but with so many tort lawyers around to apply the brakes, the bigger question is this: Will humanoid machines enrich our social lives, or will they be a new kind of television, destroying our relationships with real humans?' – Science and technology writer Fred Hapgood in 2008.

'In a properly automated and educated world, then, machines may prove to be the true humanizing influence. It may be that machines will do the work that makes life possible and that human beings will do all the other things that make life pleasant and worthwhile.' – Classic SF writer Isaac Asimov.

Chapter 2

Milestones

In Fact and Fiction

BC Dates

2000 BC. The Egyptians construct **simple mechanised toys**, such as crocodiles, made out of wood and clay. Some are on display in the East Berlin Museum, the British Museum in London, and the New York Metropolitan Museum. (Some anthropologists believe these 'toys' to be funerary objects placed in tombs to protect people in their next lives.)

322 BC. The Greek philosopher **Aristotle** expresses the role of robots thus: 'If every tool, when ordered, or even of its own accord, could do the work that befits it… then there would be no need either of apprentices for the master workers or of slaves for the lords.'

200 BC. The Greek inventor and physicist **Tesibius of Alexandria** designs **water clocks** incorporating movable figures. Water clocks are a big breakthrough for timepieces. Until this point, the Greeks have used hour-glasses which need to be turned over after all the sand has run through. Tesibius's invention measures time as a result of the force of water falling through it at a constant rate.

100 BC. **Hero of Alexandria** designs **pressure-operated mechanisms** for opening and closing temple doors. Heated air forces water into a suspended bucket, pulling on ropes wrapped around rotating cylinders attached to the door hinges. Weights cause the doors to close as water drains from the bucket.

AD Dates

AD 1350. The cathedral of the city of **Strasbourg**, in eastern France, sports a mechanical rooster. Every day at noon it 'crows', albeit silently, and flaps its wings. It is made of iron, copper and wood, and is operated by rods and levers connected to a clock mechanism. (Considered the

6 Famous Robots & Cyborgs

AD 1470s. oldest preserved automaton, the rooster can still be seen today in Strasbourg's Museum of Decorative Arts.)

In Nuremburg, the scholar Johann Müller, known as **Regiomontanus**, supposedly creates the famous '**wooden eagle of Regiomontanus**', which flies from the city of Koenigsberg to meet the emperor, salutes him, and returns. He also built an iron fly, said to fly out of Regiomontanus's hands at a feast and return to him.

AD 1495. **Leonardo Da Vinci** designs his '**anthrobot**', a mechanical man. The result of his studies of human anatomy, it is intended to mimic the movements and appearance of a knight in a fifteenth century suit of armour. It is designed to sit up, wave, and move its head while opening and closing its jaw. This robot design will influence Da Vinci's later studies of anatomy, in which he models the human limbs with cords to simulate the tendons and muscles.

AD 1590. In **Spenser's** epic *The Faerie Queene*, there is a 'metal man' called **Talus** who pursues villains relentlessly and never tires. He runs around the island of Crete armed with an iron flail.

AD 1738. The **automata** of the Grenoble-based inventor **Jacques de Vaucanson**, which include a flute player, a second automaton which can also play a drum and tambourine, and the famous **Vaucanson's Duck**, a robotic duck with each wing containing 400 moving parts, and which can apparently 'digest' grain. Detailed plans do not exist, and to this day Vaucanson's Duck remains something of a mystery.

AD 1760s. The famous **Jacquet-Droz automata** are built: the Musician, the Artist and the Writer. Among the earliest anthropomorphic robots, they have realistic faces and clothes, and can perform the simple tasks of writing, drawing and playing an organ.

AD 1770. **The Turk**, supposedly a chess-playing automaton which could defeat a human player, begins to be exhibited around Europe. Created by Wolfgang von Kempelen, it is exposed in the 1820s as a hoax. The complex arrangement of cogs and other machinery inside the Turk's desk hides a human operator, so it doesn't count as a true robot.

AD 1816. In E.T.A. Hoffmann's story, *Der Sandmann*, the hero Nathanael is captivated by the beautiful and bewitching **Olympia**, who turns out to be an automaton. She rather gives herself away by making ghastly noises like a seagull. The story was the first in an 1817 book of stories titled *Die Nachtstücke* (*The Night Pieces*). Also features in Offenbach's 1881 opera based on the stories.

AD 1822.	Charles Babbage, regarded as the father of modern computing, first presents a prototype of his 'Difference Engine' to the Royal Astronomical Society. His paper is entitled *Note on the application of machinery to the computation of astronomical and mathematical tables*. A working model of the Engine is never built in Babbage's lifetime, but the Science Museum constructs one to Babbage's plans between 1989 and 1991.
AD 1868.	The 'Newark Steam Man' is invented. Sometimes called the 'nineteenth-century steampunk robot', he is the creation of one Zadoc Dederick from Newark, New Jersey, at the age of just 22. The Steam Man is a robot in a top hat, with an enamel face, designed to pull carriages and to look as human and un-frightening as possible. Steam is generated in his trunk, which is a 3-horse-power engine, and this drives the mechanism which moves his limbs. The inventor's ambitious plan for 'steam stallions' sadly doesn't get put into practice.
AD 1886.	*The Future Eve* by Villiers de l'Isle Adam features Hadaly, the android, which is thought to be the first use of this word. Later, the word 'gynoid' for a robotic representation of the female form is to gain currency.
AD 1897.	The George Méliès short film *Gugusse et l'automaton* shows the reactions of a clown to a robot.
AD 1898.	Nikola Tesla, the Serbian-American inventor and engineer most famous for his developments in electromagnetism, builds and demonstrates a radio-controlled model torpedo boat. He claims to have invented the art of 'telautomatics', a form of robotics. (Tesla later foresees sci-fi's ubiquitous use of the laser gun, when in his treatise *The Art of Projecting Concentrated Non-dispersive Energy through the Natural Media* ,he describes a particle beam weapon capable of stopping armies in their tracks.)
AD 1907.	Tik-Tok, the round mechanical man, first appears in L. Frank Baum's *Ozma of Oz*, and in the early motion picture *The Fairylogue and Radio-Plays* (1908). He runs on clockwork springs which need to be wound up, and is thought to be the first 'mechanical man' in twentieth-century literature.
AD 1907.	The comic film *The Mechanical Statue and the Ingenious Servant* features a statue which dances when it is wound up.
AD 1921.	First performance of Karel Čapek's play *R.U.R.* (*Rossum's Universal Robots*), the work credited with bringing the word 'robot' into the English language. It comes from the word 'robota', meaning

8 Famous Robots & Cyborgs

	'serf labour', and by extension, 'drudgery' or 'hard work'. The play concerns the creation of artificial humans in a factory; they seem happy to work at first, but then think for themselves and rebellion ensues.
AD 1921.	The 80-minute Italian film *The Mechanical Man*, directed by actor/director André Deed, brings audiences a man-shaped, mechanical device used for nefarious purposes by a criminal gang. This will be echoed over 50 years later in Terrance Dicks's *Doctor Who* script *Robot*.
AD 1923.	E.V. Odle's *The Clockwork Man* depicts another of the twentieth century's first mechanical men in fiction. This is Odle's only novel, and is today regarded as something of a lost classic.
AD 1926.	Perhaps cinema's first memorable 'gynoid' or female robot, Maria/Futura, the **'Maschinenmensch'**, appears in Fritz Lang's *Metropolis*. Later robots to owe a lot to the sleek metallic design of the 'Maschinenmensch' include C-3PO from *Star Wars* and Andrew from *The Bicentennial Man*.
AD 1930s.	**Fictional robots** now start to turn up all over the place with a vengeance, e.g. in the works of David Keller, Harl Vincent and John Wyndham. These authors are tapping into a fear of this new idea of the 'robot' which is now entering the human consciousness – a fear of technology and dehumanisation, perhaps exacerbated by the uncertainties of the Great Depression. One high-profile aspect of this backlash against the 'mechanised' is the **American Federation of Musicians'** outcry about 'mechanical music' replacing live musicians in the theatre… *plus ça change* ?
AD 1936.	Cinema's first robot foot-soldiers appear – the evil Ming The Merciless's **Annihilants** in *Flash Gordon*. Sadly, as shock-troops go, they are not all that memorable and are rather easily defeated by our hero. Only slightly more menacing, from the same year, are the Volkite Robots from the Republic serial *Undersea Kingdom*, which look like a cross between the Tin Man and gun-wielding boilers.
AD 1936.	Mathematician Alan Turing comes up with the **'Turing Machine'**, a theoretical device which simulates the logic of computer algorithms.
AD 1937.	A basic robot called **Elektro** is built by the Westinghouse Electric Corporation. It can respond to voice commands, and has a vocabulary of about 700 words (recorded on 78-rpm vinyl record). He is exhibited at the New York World's Fair in 1939. He is tall, chunky, gleaming, square-headed, and actually rather fearsome-looking – like a sci-fi robot come to life.

Milestones 9

AD 1930s/40s. Lilliput, the first mass-produced robot toy, is produced in Japan. It's made of tinplate, is squarish and yellow, about 15cms tall, and runs on clockwork. Toy experts still debate over the exact date when this first came on to the market. Lilliput has a classic, blocky robot look which is much imitated in the future.

AD 1940s. The appearance of **Gnut** in *Farewell to the Master* by **Harry Bates**, who later becomes Gort in *The Day The Earth Stood Still*.

AD 1941. Isaac Asimov's many robots start to appear on the scene (including his famous robot detective **R. Daneel Olivaw** in 1954). Asimov's Three Laws first appear in the 1942 short story *Runaround*. He is credited with the popularisation of the term 'robotics.'

AD 1948. Norbert Wiener's *Cybernetics* is published. The word 'cybernetics' is a neologism coined by Wiener to describe the study of control and communication in the animal and the machine.

AD 1954. American inventor **George Devol** coins the term '**Universal Automation**' and invents the first programmable robot. This leads to the first industrial robot, Unimate, in 1961, which works on a General Motors assembly line in New Jersey.

AD 1959. Iconic robots in written sci-fi include Fritz Leiber's self-employed robot writer **Zane Gort** in his futuristic satire of the literary world, *The Silver Eggheads*.

AD 1963. First appearance of the **Daleks** in *Doctor Who*, TV's most famous cyborgs. Their gliding motion is partly inspired by the dancers from the Georgian National Ballet, whose performance in long skirts hides their feet, meaning that they seemed to glide across the stage.

AD 1965. Jean-Luc Godard's film *Alphaville,* featuring the dictatorial computer **Alpha 60**, which outlaws concepts like love, poetry and emotion. It is finally destroyed by programmer Natacha von Braun's discovery of the concept of individual human love.

AD 1967. **Surveyor III**, the third lander from the USA's unmanned Surveyor programme, is the first robot manipulator on the moon. It lands on the moon on 20 April 1967. It features an extendable arm with a scoop attachment for taking soil samples. Sadly 'dies' on the moon when it cannot be reactivated after a shutdown.

AD 1968. At the MIT (Massachusetts Institute of Technology) Artificial Intelligence Lab. Marvin **Minsky's Tentacle Arm** is developed. It is a robotic arm with 12 joints and can be operated by computer or joystick.

AD 1968. The publication of Philip K. Dick' s *Do Androids Dream of Electric Sheep?* the book which inspired the film *Blade Runner*. It brings us the concept of androids so advanced they believe themselves to be human, and has everyone who reads it nervously eyeing up their work colleagues.

AD 1969-72. The **Stanford Hydraulic Arm**, the first electrically-powered, computer-controlled robot arm, is designed by Victor Scheinman and developed at Stanford University. Its six degrees of freedom mean that it is capable of moving in all directions.

AD 1971. Writer **Stanislaw Lem** brings us the concept of '**Personoids**', robots so advanced that they do not need physical form and can live inside a computer – predating by more than a decade, all those 1980s narratives obsessed with virtual consciousness and computers with personalities.

AD 1972. Robots meet the 'blue movie', in X-rated 'erotic spoof' *Flesh Gordon*. (Time has not been not kind to the porn-fuelled satire, which is described today in *The Encyclopaedia of Fantastic Film and Television* as 'juvenile and almost defiantly grotty…the least erotic sex film ever.')

AD 1976. Robot arms are used on the **Viking 1 and 2 space probes**. The 'sampler arm' is used to gather samples from the Martian surface, and sports a temperature sensor and a magnet.

AD 1978. It's worth mentioning that the galaxy-conquering cult of *Star Wars*, as well as spawning the obvious action figures of R2-D2, C-3PO, Darth Vader, etc., also brings some of the film's more obscure characters to closer public attention through its tie-in toy line – among them the squat, unprepossessing Power Droid, basically a blue box on flexible legs. Sadly, the toy has no cool weapons or accessories and really just… stands there. Completists, of course, still need to own it.

AD 1979. The **Stanford Cart**, a wheeled robot invented in the early 1960s, but which has sat unused in a lab for most of the decade, is the focus of a new experiment. It crosses a room without human assistance, negotiating the obstacles of chairs in the way. The cart has a TV camera mounted on a rail, which takes pictures from multiple angles and relays them to a computer. The computer analyses the distance between the cart and the obstacles.

AD 1981. As real life catches up with science-fiction, **computers** start to appear in every home, with Clive Sinclair's ZX-81 proving enormously popular. The home computing revolution goes on to inspire films like *Electric Dreams* and *War Games*, giving hope to every teenager who ever dreamed of having a fantasy life through their ZX Spectrum. Sadly, Facebook is still two decades away.

Milestones 11

AD 1980s. Robots begin to be used in earnest in industry, bringing key developments such as the **Remote Reconnaissance Vehicle** and the **CoreSampler**, both used to investigate the Three Mile island nuclear accident of 1979; the outdoor navigation robot **Terregator** and the **Remote Work Vehicle** used in clean-up operations.

AD 1984. **The Terminator**, played by Arnold Schwarzenegger, becomes one of cinema's most memorable mechanical killing machines as futuristic dystopia and mistrust of machines again become popular themes.

AD 1984. One of the most extraordinarily over-priced robot toys ever, **Maxx Steele**, hits the toy market. At over $500, it sports a hefty price-tag for a toy – although its fans claim it is rather more than that. Maxx is the leader robot of Robo Force, a toy line created by Ideal.

AD 1985. **R.O.B.**, the **Nintendo Robot**, comes out as an accessory for the Nintendo Home Entertainment System. It says a lot that this particular toy robot – a little square silvery-grey chap with arms and legs – is now almost totally forgotten. It is released as a way of branching out and anticipating the crash in the video game market, but for some reason it never really catches on.

AD 1990s. Popular robots include the animated **Bender** from *Futurama*, the shape shifting **T-1000** from the *Terminator* franchise, the '**fembots**' of *Austin Powers* and the return of the **droids** for the new, if disappointing, *Star Wars* prequels. In the world of fictional cyborgs, the **Borg** from *Star Trek: The Next Generation* make their presence felt.

AD 1991. Hapless time-travellers **Bill** and **Ted** contend with their own **robot doubles** in *Bill & Ted's Bogus Journey*. The evil robots kidnap their originals and fling them to their deaths, forcing them into a battle with the Grim Reaper.

AD 1998. The **Furby** appears in toyshops and is an instant hit. This small, furry, toy animal with large eyes and a beak-like mouth, contains electronics and sensors which make it an animatronic 'pet', capable of uttering 800 words. The sinister objects, produced by Tiger Electronics, are obviously designed to frighten dolls and teddy bears everywhere. The only surprising thing is that they don't have their own spin-off TV show (although one 45-minute TV special, *Furby Island*, is made).

AD 1998. Robot mayhem comes to TV studios with the advent of *Robot Wars*, the ultimate gameshow for the geek with the killer instinct. Deadly robots are pitted against one another each week in an orgy of mechanised gladiatorial combat.

12 Famous Robots & Cyborgs

AD 1999. First appearance of **AIBO**, the Japanese robot dog, commercially available as an automated pet. Yes, now you too can have your very own K-9 around the house. AIBO is considered to have a degree of autonomy, as he can learn and mature from commands given by his human owner. He's discontinued in 2006.

AD 2000s. Film and TV remakes abound, with the **Cylons and the Transformers**, among others, given twenty-first-century makeovers as the writers and directors who grew up with these iconic robots revisit their childhood.

AD 2003. Morgui, a skull-faced robot with five senses, developed by the University of Reading, is deemed 'too scary' to be shown to under-18s. Morgui, whose name means 'Magic Ghost' in Mandarin, is an experiment in how people react to robots. The cadaverous metal head is created by Professor Kevin Warwick of the University of Reading, who says, 'We want to investigate how people react when they first encounter Mo, as we lovingly like to call the robot.'

AD 2005. An entire cast of Robots features in the film *Robots*, a mixture of cuteness and crash-bang-wallop action. It's a computer-animated film produced for 20th Century Fox, and featuring a hero called Rodney, robot son of Herb Copperbottom, dishwasher at Gunk's Greasy Spoon Diner. Rodney invents a 'Wonderbot', which sets him off on a journey to find his destiny at Bigweld Industries.

AD 2010. Inventor **Hiroshi Ishiguro** unveils the **Telenoid**, designed as a robot companion for those unable to be with their loved ones (an operator controls it via webcam and the Telenoid mimics its movements). However, its truncated limbs and foetus-like appearance are unsettling, to say the least.

AD 2010. **HRP-4C**, also from Japan, is the world's first fully-automated pop star. Only her oddly large hands and armoured silver legs give her away. She sings and dances with disturbing conviction. It can't be too long before one of these auditions for *The X-Factor* or *Britain's Got Talent* ... and wins.

AD 2011. Automated 'poop-scooper' robot **PR2** is developed by the University of Pennsylvania. It can seek out, identify and collect dog-mess. It's a chunky, unwieldy-looking creature, though, in contrast to those automated pop-stars, so maybe some sort of job-share is called for?

AD 2012 and beyond. The future will being us the culmination of **Project Aiko**, the most lifelike robot yet. Her name means 'love-child'. She isn't exactly indistinguishable from a human being and wouldn't pass a Turing Test, but... Her inventor Le Trung is attempting to reproduce a number of, ahem, 'functions'. Controversy-loving tabloid the *Daily Mail* has already got hold of this one, headlining the story 'Computer geek too busy for romance builds himself the perfect girlfriend.' Aiko can solve maths problems and has a 13,000-sentence vocabulary in both Japanese and English. She is also apparently 'the first android to react to physical stimuli and mimic pain', and this technology could be applied to people with amputations. We are told that 'Aiko is the first step towards a life-like mechanical limb that has the ability to feel physical sensations.' *Blade Runner* Replicants (crossed with *Austin Powers* Fembots) here we come?...

However, before we get too carried away looking into the actual, possible future of the automaton, let us explore the wonders of its rich and varied history, in both fact and fiction – in the form of the **FAMOUS ROBOTS AND CYBORGS**.

We'll be looking at them under the following key headings.

Also known as: Any other 'noms de robot' under which they were known.

Key Narratives: The film, TV series, book or play for which they are best known.

Appearance: Just a little about the way they look to our human eyes.

Design classic? Iconic or just an eyesore?

Origins: Who made them, when and why?

Designation: What's their role in life?

Weaponry: Just what have they got tucked away for defence and attack, if anything?

Fear Factor: ☻ up to ☻ ☻ ☻ ☻ ☻ Just how much does the metal meanie make you want to hide behind the sofa or shrink into your cinema seat?

Cuteness Factor: ♥ up to ♥♥♥♥♥ depending on how much you want to hug the metal scamp/ fearsome battle droid/ lissom replicant.

Artificial Intelligence: Could they cut it with a mechanical MENSA? Are they showing signs of being able to think for themselves?

Skills: What can our robot or cyborg actually do? Does it have any special abilities?

Catchphrase: What's the one line everybody quotes?

Notability: ✯ up to ✯✯✯✯✯ How well-remembered is the robot/cyborg?

Databank: Some key facts.

Quotes: Anything interesting which has been said by or about our robot/cyborg.

Further Appearances: Also seen in…

Nagging Questions: What doesn't make sense?

Memorable Moment: One stand-out scene in our robot or cyborg's life.

Assessment: A brief rundown on the impression they leave us with.

Verdict: The final judgement.

Chapter 3

The First Ones

OLYMPIA

Also known as:

Key Narratives: *The Sandman* (*Der Sandmann*) by E.T.A. Hoffmann, the first in an 1817 book of stories titled *Die Nachtstücke* (*The Night Pieces*).

Appearance: Human female.

Origins: Created by Professor Spalanzani and the mysterious Coppelius.

Designation: Automaton, driven by clockwork.

Weaponry: n/a

Fear Factor: 👽 👽

Cuteness Factor: ♥♥ (For her outward beauty, but then belied by her inner grotesqueness.)

Artificial Intelligence: Seemingly quite high functional intelligence and movement, but still only capable of 'AH, AH!' noises.

Notability: ✯✯

Databank
- The story begins with the memories of the young Nathanael who recalls vividly the nightmare of the 'Sandman' who removes children's eyes. He associates him with the lawyer Coppelius, a dinner guest of his parents, who apparently threatens to put hot coals in the boy's eyes when he catches him eavesdropping – although Hoffmann leaves it ambiguous as to how much of this is in the boy's imagination.
- Later in life, the character of Coppelius becomes associated in Nathanael's mind with one Giuseppe Coppola, an itinerant pedlar, from whom he buys a spyglass. This enables him to see into the house of the Professor, where he glimpses the beautiful, but oddly silent and motionless Olympia. His love for her becomes a terrible, compelling fascination and he even forgets the sweet Clara to whom he is engaged.

- The appearance of Olympia in the story is as much satirical as it is grotesque – the author is intending to make fun of the way the hero Nathanael idealises his beloved Clara and, by extension, the idealising of woman by man.
- Moira Shearer plays Olympia in the 1951 film version of *The Tales of Hoffmann*. While dancing, she mimes to the voice of singer Dorothy Bond, although there are occasional lapses in the miming.

Quotes
'Nathanael dug up everything he had ever written … and all of this he read to Olympia tirelessly for hours at a time. Never before he had such a splendid listener. … she sat for hours on end without moving, staring directly into his eyes, and her gaze grew ever more ardent and animated.'

Further Appearances
The story was loosely adapted as the 1870 ballet *Coppélia*, and also as part of the opera *The Tales of Hoffmann* by Offenbach.

Memorable Moment
Nathanael dancing with Olympia. He is so obsessed with her that he is unable to see beyond external beauty and even to find it strange that her repeated 'Ah! Ah!' noises do not seem remotely human. There is something comically ridiculous, but also tragic and disturbing about this scene.

Assessment
One of the earliest automata in fiction appears in a narrative which is not a futuristic story 'about' humanity's interaction with robots, but rather a contemporary story of the 'uncanny' with strong satirical and psychological undercurrents.

Verdict: Living Doll.

HADALY

Key Narratives: *The Future Eve*, a novel by French symbolist writer Villiers de l'Isle Adam, published in 1886.

Appearance: Human female, based on the protagonist Lord Ewald's beloved, Alicia Clary.

Origins: Constructed by a fictionalised Thomas Edison.

Designation: Physically perfect android replica.

Weaponry: n/a

Fear Factor: ☻

Cuteness Factor: ♥♥

Artificial Intelligence: Unclear.

Notability: ✯

Databank
- Edison's experiments are carried out in a secluded estate outside New York City and an underground cave full of inventions.
- In the creation of the android, Edison is seen to use techniques which were cutting-edge for the nineteenth century: photosculpture, photography and the phonograph.
- Once Alicia's likeness has been artificially reproduced on the form of an android named Hadaly, (the name, according to the author, means Ideal in Persian), she is presented to Lord Ewald, who at first does not even realise that she is an android and not the original Alicia.
- Ewald and Hadaly board a ship, the *Wonderful*, to make their home back in England, but it sinks and Hadaly is consigned to the bottom of the ocean in her travelling coffin.
- The story is notable as the first text to bring us the word 'android', now defined as a robot, which is designed with the appearance and behaviour of a human – often, in science fiction, with the intention of impersonating or supplanting a particular human. Android doubles have been a staple of imaginative fiction and feature in everything from *Doctor Who* to the *Bill and Ted* films.
- With its themes of the replication of the human form, exploration of what it means to be human, and creating the perfect woman, it anticipates *The Stepford Wives* and classic 1980s sci-fi like *Blade Runner* and *Weird Science*.
- Both *Der Sandmann* and *The Future Eve* use the figure of the mechanised woman to show us how ineffective a technological reproduction of humanity can be. Despite presenting an outward, 'perfect' facsimile of humanity to the world, neither Hadaly nor Olympia possesses the traits which make Alicia and Clara the women that they are. The facets which the reproductions have eliminated – spark, spirit, independence, wilfulness – might ostensibly irk their lovers, but are in fact, the qualities which make them more than mere automata.
- Both texts are satirising the social etiquette of the nineteenth-century bourgeoisie, in which mechanised surface action, 'enacting' particular social codes and rhythms, has become more important than deeper human understanding, spontaneity.

Assessment
A controversial work, accused of misogyny because of its presentation of the female form as object and fantasy wish-fulfilment. Hadaly is intended to replace Alicia by emulating all of her physical beauty but none of the 'bothersome' aspects of her

personality! The book also deals with the relationship between technology and the arts, and the viewing of the female as both subject and object – two ways in which it can be seen as ahead of its time. Hadaly is one of the first characters in literature to make the reader ask questions about 'artificial intelligence' centuries before that term officially existed.

Verdict: An early dip into the world of A.I., forcing us to ask some interesting questions.

THE TIN MAN

Also known as: The Tin Woodman/Tin Woodsman.

Key Narratives: *The Wonderful Wizard of Oz,* 1900 novel by L. Frank Baum, and its best-known adaptation, *The Wizard of Oz,* 1939 film.

Appearance: A man made out of tin, with a massive chopper. He has largely metallic parts, but his face is pliable, like that of a human. Only scarily silver.

Design classic? Holds up surprisingly well, and one fears for how he might end up looking in a remake…

Origins: (Re) built by a tinsmith. See below.

Designation: Law-abiding cyborg (technically).

Artificial Intelligence: States that he has no brain, but is not especially bothered about losing it.

Weaponry: Large Chopper.

Fear Factor: 👁 👁 (The actors playing the Tin Man, Scarecrow and Cowardly Lion, were asked to eat in their dressing-rooms, rather than the studio dining hall, in case their costumes scared people… and there is something pretty unsettling about the Tin Man's silvery, apparently human, face underneath all that metal…)

Cuteness Factor: ♥♥♥ (Having said that, his desire for a heart must touch the sensibilities of some…)

Skills: Wood-chopping and breaking into song-and-dance at tiresomely regular intervals.

Notability: ★★

Databank

- In the book, he is a character in the Land of Oz called Nick Chopper, whose enchanted axe somewhat gruesomely removed his limbs. Like the Cybermen and

the Six Million Dollar Man, he was rebuilt, replacing his limbs with tin each time he lost one. (History does not report how he dealt with the attendant blood loss and presumable threat of gangrene and septicaemia, or why he simply didn't die of shock, but we'll let that one pass.)
- Buddy Ebsen, who was going to play the part, inhaled aluminium powder from the make-up and became ill, so Jack Haley played the role in the end.
- The costume was so heavy and inflexible that Haley needed to lean against a board if he wanted a rest and he could not sit or bend down in it. He didn't find making the film an enjoyable experience. The costume no longer exists and is thought to have been destroyed.
- The 'Nick Chopper' backstory is all lost in the film. It is simply implied that the Tin Man has always been made of tin. The only reference to the tinsmith is the line 'The tinsmith forgot to give me a heart', which seems like either (a) a remarkable piece of shoddy tinsmithery or (b) a conspiracy. So if we take the film as our main narrative, the Tin Man can perhaps be classed as someone with aspirations to being a cyborg, but who isn't one yet…

Further Appearances
Further Oz books by L. Frank Baum, most notably *The Tin Woodman of Oz* (1918). See also the Gregory Maguire novel *Wicked: The Life and Times of the Wicked Witch of the West* (1995) and musical adaptation *Wicked*. There's a 1961 animated TV series *Tales of the Wizard of Oz* in which the Tin Man is (amusingly or annoyingly – you decide) renamed Rusty, and is voiced by Canadian actor/ disc jockey Larry D. Mann.

Nagging Questions/Memorable Moment
The Tin Man finally gets the heart he has been after all this time and, um, it's a cushion stuffed with sawdust. Or in the film, a clock. Thanks. Most people would find this a tad disappointing, but he seems to take it all in his metal stride – because it's all terribly deep and meaningful and symbolic, you see, and he has 'heart' anyway without the need for any physical entity pumping blood around his tin body. (Which, let's be honest, probably wouldn't work anyway without some extremely complex and messy cybernetic surgery.) It's all to do with how much he is loved by others. Dodgy science, but acceptable cinematic cod-philosophy.

Assessment
The Tin Man's desire for a heart seems rather odd, as he seems to manage perfectly well without one – and, indeed, when the Wizard is proved to be a great big fake at the end (spoilers, sorry!) he is unable to provide him with a proper one. Perhaps one should not be applying logic here. Given the irritating status of the Oz sequences in the film as one long, elaborate fantasy – the ending is, literally, that old Year 5 writing exercise and *Dallas* Season Eight trope, 'And then I woke up and it was all a dream' – the very existence of the Tin Man at all as a character in any fictional universe

outside Dorothy's fevered imaginings is, at best, debatable. (See also *Chitty-Chitty Bang Bang,* where a large part of the narrative is simply a story told by Mr Potts to the children, including their adventure in the Baron's castle – and so by the end, in real time, he has only just met Truly Scrumptious and enters into a somewhat implausible romantic liaison with someone who is practically a stranger.) Anyway, the Tin Man is yet another iconic character, who everyone will surely remember having been exposed to at some point by one of the film's perennial TV outings at Christmas and other Bank Holidays. It is possibly the most-watched film in the Western World.

Verdict: He showed his mettle.

ROBOT MARIA

Also known as: *Maschinenmensch* or man-machine.

Key Narratives: The film *Metropolis* (1927) directed by Fritz Lang, and based on Thea von Harbou's 1925 novella.

Appearance: Gleaming, sleek, metallic android (technically, 'gynoid'), obviously modelled on the female form, and then an exact double of Maria. The script refers to the appearance of an Egyptian statue.

Design classic? Undoubtedly.

Origins: Created by the scientist C.A. Rotwang, played by Rudolf Klein-Rogge. It's originally intended as a replacement for his wife, Hel, who left him for the Master of the City, Fredersen, (she died while giving birth to Fredersen's son, Freder). Rotwang then uses the robot in his grand scheme for revenge, while pretending that he is using the robot to aid Fredersen…

Designation: Machine-being.

Weaponry: n/a

Fear Factor: 👽 👽

Cuteness Factor: ♥♥♥ (Probably depends how much you fancy the actress Brigitte Helm.)

Artificial Intelligence: Present.

Skills: spreading dissent…

Notability: ✯✯✯✯✯

Databank

- The transformation of Machine-man / *Maschinenmensch* into the double of Maria is one of the most memorable scenes in the silent film *Metropolis*. It's where a new sci-fi sensibility meets the Gothic, taking place in a laboratory where the trappings of science fiction – lamps, crackling sparks, switches, a sleek glass coffin-like unit – overlap with those which in previous centuries would signify the 'dark arts', like bubbling flasks of dark liquid and swirling clouds of smoke, and Rotwang's appearance as an unhinged necromancer-like figure.
- Rotwang lives in a weird house in the middle of Metropolis, oddly rough-looking and ramshackle compared to the sleek, elegant city. It has a trapdoor leading to catacombs, where Rotwang and Fredersen eavesdrop on a secret meeting of the workers and Maria.
- Rotwang lost a hand while developing the Machine-Man and now wears a prosthesis in its place – an early example, perhaps, of cinema's association of cybernetics with the forces of evil.
- Rotwang is asked to create the android Maria as a means of spreading discord between her and the workers to whom she preaches.
- During the riots and power blackout, Rotwang goes mad and chases Maria through Metropolis in the belief that she is Hel. Freder pursues him to the roof of the Cathedral and the two battle in a manner reminiscent of Moriarty and Holmes on top of the Reichenbach Falls, and Rotwang falls to his death.
- The robot was designed by one Walter Schulze-Mittendorff, whose original plan to make the costume out of beaten copper was abandoned as it would make it too heavy and uncomfortable for the actress inside. In the end he opted for a new substance, 'plastic wood', coated with cellon varnish spray mixed in with a silvery-bronze powder. He then created the costume in sections, like armour, around a plaster-cast of the 17-year-old actress Brigitte Helm.
- The cast was made standing up, and so the costume proved constricting for the actress – simple actions like sitting down proved difficult – and she was plagued by scratchy, sharp edges inside it.

Quotes

'What excited me most about the role of Maria in Metropolis were the character's crass differences, because these also lie hidden in my own nature: the austere, pure and chaste Maria, who believes in doing good, and the Maria the obsessed siren. Whenever I'm told how well I portrayed these intertwining and contradicting elements, I find it flattering and take it as a compliment. It was incredible work. Now that it's over, I have trouble remembering the disheartening and sadder moments – only the sunnier and uplifting moments stay with me. Sometimes it was like heaven, and other times like hell! The three weeks spent shooting the water sequence, when the underground city is flooded, were

unbelievably hard on my health. Even now, I have to admit that I don't know how I got through it.' – Brigitte Helm, writing in 1927.

Further Appearances
There is a 1949 manga version of *Metropolis*, loosely based on the film and published in English by Dark Horse. This in turn inspired a 2001 animated film version. The original film was re-released in 2002 and restored with extra footage in 2010.

Memorable Moment
The effects are splendid – especially notable when the *Maschinenmensch* is surrounded by halos of energy as it takes human form beneath a helmet-like object. Decades later, there is still discussion – and no definite consensus – as to how this effect was achieved, and it set the bar for special effects in the decades that followed.

Assessment
Metropolis seems decades ahead of its time with its effects, cybernetics, class-war themes and warnings about the (almost erotic) fetishising of technology. On its re-release in 2002, the *San Francisco Chronicle* described it as *'a time-bending experience, a way of visiting the past and glimpsing the past's idea of the future. A masterpiece of art direction, the movie has influenced our vision of the future ever since, with its imposing white monoliths and starched façades.'*

Verdict: The tensions between (wo)man and machine, Expressionistically expressed.

THE ANNIHILANTS

Key Narratives: *Flash Gordon* film serials 1936-40.

Appearance: Tall, chunky, square-headed, with a central chest-panel and wearing metallic smock-like garments, they look less than fearsome by today's standards – like a cross between the Tin Man and a child's *Blue Peter* model of a Cyberman in a silly hat.

Design Classic? Um, how can we put this? No. Can't even see it catching on in a retro manner.

Origins: Creations of Ming the Merciless from the planet Mongo. Yes, honestly.

Designation: Remote-controlled, robotic troops.

Weaponry: Weapons of destruction – walking bombs which require human control. The flaw in this plan can probably be seen very quickly.

Fear Factor: ☻ (Despite trying really, really hard – sorry.)

Cuteness Factor: ♥

Artificial Intelligence: Low, pretty much non-existent.

Skills: Strength. Not many more… apart from getting blown up.

Notability: ✯

Databank
- The first in a glorious line of cackling movie villains with convoluted plans who inexplicably rely on the performance of unreliable or easily-defeated henchmen, Ming The Merciless perhaps somewhat undermines his own credibility – and, indeed, his scary moniker – by using the Annihilants. Laughably ineffective and lumbering, they don't even have lasers.
- They are designed as 'perfect' soldiers to conquer and destroy humanity. They are built up as being invincible because of their great size and strength, but are ultimately destroyed with great ease. Um, back to the drawing-board, Ming.
- Ming's later schemes for controlling the universe, interestingly, all involve the 'automation' of humanity in some way – either through drugs or charms or futuristic weaponry. He obviously has something of a fetish for control, one which extends beyond the mere building of robotic troops.
- Here, as in many later sci-fi films, robotic control/technology is seen as the preserve of villainy – not evil in and of itself, but easily exploited and misused. The organic power of humanity is seen as the force for good, out-thinking the logical processes of robotics. Human ingenuity, individuality and flair are positive traits which Flash embodies, as do later classic sci-fi heroes from Luke Skywalker to Sarah Connor.
- For more on this, see J.P. Telotte's *Replications: A Robotic History of the Science Fiction Film* (University of Illinois, 1995).

Memorable Moment
The Annihilants are designed to be walking bombs, and Flash Gordon destroys them – by blowing them up. One hesitates to point out the flaw in the plan of an intergalactic conqueror, but…

Assessment
As an early example of cinematic robots, the Annihilants compare poorly with their cousins from a decade earlier. They are a step back in scariness terms from the Mechanical Man of the 1921 silent movie, and lack any of the sleek grace of Maria from *Metropolis*. They are interesting, though, from two points of view: they remind us that a robot's effectiveness is dependent on its programming and operation by human beings (we are a long way off 'artificial intelligence' here), and they are an early example in cinema of an army of automated servants with a human

commander, a template which would be replicated in many texts throughout the twentieth century.

Verdict: Annihilated. Um, rather easily.

GORT

Key Narrative: *The Day The Earth Stood Still* (1951), remade 2008.

Appearance: Minimalist, bipedal helmeted warrior.

Design Classic? Yes. Sleekly beautiful and menacing.

Origins: Klaatu's planet, as a kind of intergalactic policeman.

Designation: Robot – Protector.

Weaponry: Beam weapon fired from the head-visor (1951), cloud of nanomachines (2008).

Fear Factor: 👽 👽 👽 (Even though he doesn't do much apart from just stand there, he does do so pretty menacingly.)

Cuteness Factor: ♥♥

Artificial Intelligence: Unknown.

Skills: Defence and attack, and menacingly standing on the spaceship ramp.

Catchphrase: The key phrase 'Klaatu barada nikto' is the one on which the film's decisive moment pivots.

Notability: ★★★

Databank
- A flying saucer lands in Washington DC, and the humanoid Klaatu, played by Michael Rennie in the original film and Keanu Reeves in the remake, steps out to announce he has come to make peace with Humanity. He's accompanied by Gort, who vaporises the welcoming committee's weapons following a misunderstanding.
- Klaatu's 'demonstration of power', in which he cuts electrical power to the world, is taken to be an aggressive act, and when the situation escalates, Gort is on the point of destroying the Earth until he is stopped by a phrase Klaatu has entrusted to the character Helen Benson, played by Patricia Neal. The phrase, uttered by Neal at the climax of the film, has passed into cinema legend: 'Klaatu barada nikto!' No official translation of the phrase has ever been offered, but one can guess, from its context and its effect, that it is part of some kind of 'fail-safe' procedure.

- The film, based on Harry Bates's short story *Farewell to the Master*, has come to be regarded as a classic of 1950s sci-fi. It was advertised with a beautifully melodramatic image of Gort scooping up a stereotypical 'woman in peril' – presumably a very loose depiction of Patricia Neal – while beams surge from his eyes; in the background a giant hand clutches the Earth. The film's tagline was 'From out of space… a warning and an ultimatum!'
- *Halliwell's Film Guide* describes the film as 'Cold-war wish fulfilment fantasy, impressive rather than exciting…'
- Gort is played in the 1951 film by Lock Martin, a tall actor (over 7 feet, although reports of his actual height vary). He died in 1959, aged just 42. The metallic-looking suit he wore for the part was made of foam rubber, given additional height by built-up shoes and a helmet which rose above the level of his head.
- There were two costumes, in fact, and an inert 9ft fibreglass statue which was used for some shots. Both costumes and the statue gained their silvery sheen by being painted with rubberised silver paint.
- Lock Martin was unable to pick up Patricia Neal unaided and had to be assisted by wires (in shots from the back where he's carrying her, it's actually a lightweight dummy in his arms). He also had difficulty with the heavy robot suit and could only stay in it for 30 minutes at a time. By the end of filming, the discomfort of wearing the suit was causing him to begin having spasms in his arms.
- Description in the script by Edmund North: 'After a moment a ramp slides silently out of the side of the ship and a giant robot ten feet tall steps down the ramp to the ground. He is not a metallic, clanking robot, but is made in the almost perfect image of a man. His face is, and remains, expressionless. This is GNUT.' (Gnut was the original name for the robot in the short story, and it was changed to make it easier to pronounce.)
- This description was brought to life by the art director assigned to the film, Addison Hehr. He came up with the idea of representing the robot as 'fluid metal', something with an unjointed carapace which would bend and shape itself at will. (The limitations of 1950s cinema prevented this from being realised fully. Forty years on, when watching the T-1000's fluid computerised movements in *Terminator 2*, we can perhaps have some idea of how Hehr ideally envisaged the effect…)
- The close-ups of Gort zapping its deadly ray used another prop, a larger shoulder-and-helmet unit packed with electronics. The visor was powered by a small electric motor to ensure it rose at a constant speed, and behind it was a clear, blue-tinted sheet of Lucite, an acrylic product. Behind this was a series of blue light bulbs, wired to light in a specific pattern at certain intervals for the moments in the story when Gort's visor opened. The laser 'fire' was produced by animation.

- In the 2008 film, Gort and his power are depicted entirely using CGI, and this incarnation is considerably taller than the original Gort. He's shown being able to create a destructive 'cloud' of nanomachines to devour a target.
- Seattle's Science Fiction Museum and Hall of Fame contains a Gort replica.
- Echoes of Gort's sleek, silver-suited, visored design can be seen in the look of 1987's *Robocop*.

Quotes

'Gort represents a watershed moment in science fiction ideology. Gort was a reaction to a world mired in post-Holocaust existential relativism, to belief in definable concepts of 'good and evil' and other societal and moral dictums ... The proposition that there is an absolute sense of right and wrong, or acceptable and unacceptable, is a political debate that continues to dictate peace and conflict throughout the world today.' – Don Marinelli, professor of drama and director of Carnegie Mellon University's Entertainment Technology Centre.

Memorable Moment

When Gort's visor first sends blazing rays out to evaporate the soldiers' guns – a masterful piece of special effects for the time.

Assessment

The great imaginative tradition of 1950s sci-fi epics such as *When Worlds Collide*, *Forbidden Planet* and *The War of the Worlds* dared to take audiences to unknown places, scripts and direction, infused with a new questing boldness and a dash of Cold War paranoia. *The Day the Earth Stood Still* is a classic entry in the canon. The appearance of Gort is a triumph of lateral thinking and improvisation, matching the ambition of the robot's appearance in the script. It's the minimalist simplicity of Gort which haunts the memory; the robot does little more than stand on the ramp of the flying saucer and use his visor-laser, but that, combined with the rush of almost-apocalyptic finale, is enough to ensure he is a truly memorable robot.

Verdict: Earth-stoppingly magnificent.

R. DANEEL OLIVAW

Key Narratives: The novels of Isaac Asimov, starting with *The Caves of Steel* in 1954.

Appearance: Humanoid, with high cheekbones and swept-back, bronzed hair – an idealised human.

Origins: Created by roboticists Roj Nemennuh Sarton and Dr Han Falstofe from the planet Aurora.

Designation: Robot – morally guided by the three Laws of Robotics.

Weaponry/Skills: Fast analytical processes and memory.

Fear Factor: 👽

Cuteness Factor: ♥♥♥

Artificial Intelligence: Very High.

Notability: ✯✯✯✯

Databank

- Isaac Asimov's most famous robot, and his most frequently-occurring character, who turns out to be instrumental in creating the worlds in which Asimov's fiction is set.
- First appearing in *The Caves of Steel*, the 'humaniform' Olivaw, who is indistinguishable from a human, assists Earth policeman Elijah Baley in solving interplanetary crimes. His first case is the investigation of the murder of his creator, Roj Nemennuh Sarton.
- Baley and Daneel are probably the inspiration for investigator Poul from *Doctor Who* and his robot sidekick D-84.
- Daneel ends up working in the background of human history for thousands of years, becomes a student of Humanics and formulates the Zeroth Law of Robotics – 'a robot shall never harm humanity, or through inaction allow humanity to come to harm.'
- With R. Giskard Reventlov (the 'R' in both their names indicates 'robot'), Daneel is the creator of the study of 'psychohistory', a combination of history, psychology and statistics which is used by the famous Hari Seldon in Asimov's *Foundation* series.

Assessment

One of Sci-Fi's most notable and influential automatons, and a template for a succession of fictional robots – but also a memorable character in his own right. Asimov manages to make Olivaw's adventures a running thread which ties together work from different ends of his career and contributes to its beautiful thematic unity.

Verdict: Almost human…

ROBBY THE ROBOT

Key Narrative: *Forbidden Planet* (1956 film directed by Fred M. Wilcox and written by Cyril Hume, with a story loosely inspired by Shakespeare's *The Tempest*).

Appearance: 211cm high, bipedal. Transparent domed head with two antennae, showing robotic workings inside. Two flexible arms with clamp-like hands. Bulbous pelvis/leg units each made up of four separate segments.

Origins: Built by Dr Morbius.

Designation: Robot – helpful.

Weaponry: n/a

Fear Factor: 👽 👽

Cuteness Factor: ♥♥♥♥ (Robby is one of cinema's most-loved robots, despite his initial appearance being intended to be fearsome.)

Artificial Intelligence: High.

Moral guidance: Follows the Three Laws of Robotics as set down by writer Isaac Asimov.

Notability: ★★★

Databank:
- Despite his initial somewhat frightening appearance – due in part to the unsettling combination of the bipedal, neo-humanoid form and the blank, robotic face – Robby turns out to be the moral heart of his key narrative, refusing to destroy the 'Id Monster' in the film when he realises that it is, in fact, an extension of the psyche of Dr Morbius (Walter Pidgeon).
- The 'fearsome' aspect of Robby's appearance is played up in the original movie poster for *Forbidden Planet*, which, in a classic piece of audience misdirection, presents the robot in a threatening stance and sporting what appears to be almost a malevolent, mechanical grin. Thus audiences would already be 'primed' to expect Robby to be the movie's villainous presence.
- Robby was one of the first robots in film narrative to have a 'personality' or 'artificial intelligence', and displayed a propensity for witticism – intended to reflect the fact that his personality had been programmed by Dr Morbius.
- The Robby 'suit' was built by Robert Kinoshita, following the design of art director Arthur Lonergan. The actor Frankie Darro, aged 33 at the time, operated Robby from inside the suit. The voice was provided by Grammy award-winning voice-over artist Marvin Miller.

- The components inside the robot suit needed an external power supply, and in some shots the cable leading from the suit is clearly visible.
- The 'mouth-like' arrangement in Robby's domed head, which makes such a prominent and startling appearance on the film poster, is actually a light-organ synchronised with his voice – a device designed to convert audio signals into light effects.
- Many of Robby's features can be seen to have inspired later robots, especially 'Robot' from *Lost in Space*, who shares a similar look in several respects – the glass-domed head, the rotating antennae for 'ears', the flashing mouth, the chest panel.
- Robby has been represented in toy form several times: a 1950s clockwork tin toy from Planet Robot is one of the rarer examples, and can fetch up to £20 on online auction sites.

Quotes
'He has no face — only a complicated arrangement of electronic gadgets which crackle and light-up at unexpected moments. In spite of his disproportioned arms and legs, he only very roughly suggests the human shape. His hands are tools, and various spare parts (one of these actually a metal hand) are neatly clipped to his body, back and front. He is able to rotate the upper part of his dome, and so seems to 'face' the person addressing him. A small radar antenna comes up out of Robby's dome, and slowly rotates.' – Original screenplay description.

Further appearances
The film *The Invisible Boy* (1957), three episodes of TV series *The Twilight Zone*, and an episode of *Columbo* called 'Mind Over Mayhem', among many others.

Memorable Moment
Trundling along on his 'moon-buggy', and lumbering in after being summoned by Anne Francis, Robby commands the attention of the viewer.

Assessment
One of the most famous, affectionately-remembered robots of all time, playing his part in one of the best-remembered classic sci-fi films. Recognisable simply from his silhouette, Robby has continued to delight and entertain down the decades.

Verdict: Iconic.

Chapter 4
The Best of the Worst

TV and film's most rubbish robots and cyborgs

Robots and cyborgs aren't always memorable for the right reasons. Sometimes, mechanical men come along who, for whatever reason, just don't grab the public imagination because they're too dull, too much like something that's been seen before. And sometimes, we get a mechanised monstrosity which lodges itself in the memory wafers just because it's so chillingly awful – so dreadful you want to erase it from your personal databank, or zap it into a hundred zillion sizzling particles with your own personal sizzling laser of death, so that you never have to see the damn thing again.

So here's a rundown of the most embarrassing Robots and Cyborgs of Death and Doom – bots who we hope have had the decency to shuffle off the solenoid coil and deactivate their central systems permanently. Here they are, presented in chronological order: the worst, the lamest, and the most embarrassing metal men of all time. It's the Infamous Robots and Cyborgs.

THE ANNIHILANTS in *Flash Gordon* (1936-40)
Chunky, unthreatening robot boxes which look as if they were put together by some unimaginative primary-school children – and with the fatal flaw that they are simply walking bombs. We never really buy into the Annihilants as any great threat, and they don't exactly prove a threat to Flash, who just blows them all up.

DR SATAN'S ROBOT in *The Mysterious Doctor Satan* (1940)
Supposedly the death-dealing robot henchman of the most evil villain ever. With a name like that, he obviously has delusions of grandeur. In fact, the robot looks like an animatronic metal postbox with ungainly clamp-hands, or possibly a walking hot-water boiler. Turns up again in 1949's *King of the Rocket Men*.

RO-MAN in *Robot Monster* (1953)
With a great furry body which looks as if he's pinched it off a passing Yeti – or possibly cannibalised from the many coats of Joan Collins – Ro-Man doesn't even look like a robot. His diving-helmet head with protruding antennae makes little attempt to disguise the actor within, and his lumbering gait looks distinctly like that of an awkward thesp who, just a few weeks ago, was treading the boards as Third Gentleman, and is doing this sci-fi gig to pay the rent. It's not even half-hearted and is in need of a lightning bolt up its robotic pacemaker. 'An actual preview of the devastating forces of our future', according to the enthusiastic voice-over man. We don't think so.

CHANI in *Devil Girl From Mars* (1954)
Another sci-fi epic and another depressingly inflexible robot costume, this time brought to you by director David MacDonald. Slinky vinyl-clad alien Nyah heads to Earth in search of virile males to help replace the dying men of her home planet. This kind of inter-galactic sexual predation never works out well, and it's made doubly difficult by the fact that all Nyah can do is keep coming into a country pub, make some threats and then leave again. And exacerbating the problem is her 12ft remote-control robot, Chani, a clunky thing which looks like a giant fridge surmounted with a roadmender's lamp. It's got some flexible-looking arms which don't appear to do anything, and you keep thinking that really, one good shove would deal with it for good.

THE VENUSIAN ROBOTS in *Target Earth* (1954)
There are meant to be hundreds of these things rampaging through Chicago, but budgetary constraints mean that only one robot costume was ever made for the film – and what a rubbish costume, it is too – bow legged, with American Footballer shoulders and looking as if it's been made out of a pile of cardboard boxes. While the BBC were always pretty good at making their limited resources go far, the makers of invasion movie *Target Earth* just don't seem to have got the hang of it. We need to cut older productions some slack for their lower budget and simpler technology – but when you look at the robot magnificence of *Forbidden Planet* and *The Day The Earth Stood Still* you realise what a difference a powerful script, strong performances and conviction can make.

KRONOS ROBOT in *Kronos* (1957)
This fearsome, 100ft-tall, all-conquering robot in Kurt Neumann's sci-fi romp unfortunately resembled a couple of boxes stuck together with a radio aerial shoved

on top. Maybe they were going for some kind of abstract minimalist aesthetic, coupled with frightening Modernist blankness, but it just didn't pay off. The terrifying movie poster, with KRONOS in giant fiery letters over the faces of the terrified human protagonists, promised so much...

THE HUMAN ROBOT in *The Robot vs The Aztec Mummy* (1958)
The perplexingly-named Human Robot is a boxy, *Blue Peter*-style effort, looking rather like a child's attempt to produce a scary monster in a craft session. In the film, it's designed by the evil hypnotist Dr Krupp, and engages in a fist-fight with the Aztec Mummy. Nowhere near as exciting as it may sound.

TORG in *Santa Claus Conquers The Martians* (1964)
Widely thought to be one of the worst films ever made, Nicholas Webster's *Santa Claus Conquers the Martians* is, like the films of Ed Wood, only enjoyed these days in a kitsch so-bad-it's-good way. One of the most unintentionally laughable aspects is the robot Torg, looking like a giant, ungainly toy robot with a cylindrical head. Even hand-held point-of-view shots and a stirring timpani soundtrack can't distract from the robot's essential rubbishness. And he's an anagram of the fabulous Gort, as well, which makes him doubly lame.

MECHA-KONG in *King Kong Escapes* (1967), aka Mechani-Kong
Oh, dear. Watching *King Kong*, how many people seriously thought, 'this giant gorilla is all very well as far as it goes for a movie monster, but what he really needs is an unconvincing robot duplicate to pitch himself against'? Well, in 1967, that's just what Kong gets, thanks to director Ishiro Honda and the actor in the Kong suit, Hiroshi Sekita. Created by the villainous Dr Who – yes, really, and no, not *that* one – the metallic Kong goes digging for a special radioactive element deep in the North Pole. Yes, the North Pole. He doesn't turn out to be very good at this and the evil Dr Who (I know, we can't get past that either) calls in the real Kong... Here we really want to add 'with hilarious consequences'.

BOX in *Logan's Run* (1976)
Humiliatingly rubbish from the moment he trundles himself into the ice-cave where Michael York and Jenny Agutter have fled, Box speaks like a bad Shakespearean actor and has one of the most unconvincing silver faces and boxy bodies ever to grace a movie. Calling himself 'more than man or machine', he looks as if he was put together in the special effects team's lunch-hour, frankly. The two great British thesps give their all, playing their scenes as straight as they can while wrapped in furs and seething with sexual tension, but it's a lost cause.

THE ROBOT MAID in *Come Back Mrs Noah* (1977-78)
Notoriously awful sci-fi comedy featuring Mollie Sugden, and scripted by David Croft and Jeremy Lloyd. One of the characters was a blank-faced, red-haired robot French maid – an automaton played by the lovely Vicki Michelle, adorned with a pink bow in her hair and teetering like a Weeble. Predictable gags about her 'luxury equipment' sent a screeching studio audience into inexplicable paroxysms of laughter. Although rather cringeworthy as robots go, the role was useful for her, as it led to her being chosen for the role of waitress Yvette in the same writers' rather funnier and more fondly-remembered sitcom *'Allo 'Allo.*

THE ROBOTS in *War of the Robots* (1978)
Ponderous and laughably bad 1970s sci-fi epic from Italian director Alfonso Brescia, featuring humanoid robots in silver jumpsuits and peroxide wigs, with laser guns which go *Pchow*. Now, laser guns which go *pchow* are not necessarily a very bad thing, and, given a certain level of knowing camp and directorial panache, even wigs and jumpsuits might be something you could get away with. Except here, it's just... not. The brilliant Professor Carr, purveyor of dodgy science, is kidnapped by the wiggy robots and mayhem ensues. It all ends in a grand space battle which goes on forever. Even fans of 'so bad it's good' films find this one a bit hard to endure.

H.E.R.B.I.E. from the *Fantastic Four* TV and comic series (1978-79)
This not-so-fantastic automaton – it stood for Humanoid Experimental Robot, B-type, Integrated Electronics – was an ally of the Fantastic Four in the Marvel Comics Universe, and was created by Stan Lee and Jack Kirby. They needed a replacement for the Human Torch character (as, it was dubiously claimed, children might try to emulate him by setting themselves on fire) and so the void was filled, as is so often the case, by a cute robot. Some found him loveable, but most simply found him annoying.

V'GER in *Star Trek: The Motion Picture* (1980)
Fans could not hide their disappointment with the first reunion of the original Trek cast, in which a ponderous pace, presumably intended to convey a sense of epic space grandeur, made the whole thing drag interminably. It's been given the not-totally unfair nickname of *The Motionless Picture*. The final twist (was it worth waiting for?) is that the *Enterprise* has discovered a being which has evolved out of the Voyager 6 probe, calling itself V'ger (or, in the novel, Vejur). It never really does anything. Of mild interest is that it encountered 'living machines' which gave it life, and so Trekkies have fallen on this as early evidence in the *Trek* universe of the existence of the Borg...

TWIKI in *Buck Rogers in the 25th Century* (1979-81)
Cuteness for the kids in the glossy sci-fi show, looking like a metallic Little Lord Fauntleroy, and dragging down the credibility of every scene he was in. All right, so *Buck Rogers* wasn't renowned for its tough, gritty realism, but this chirpy chappy really didn't win people over in the way he was supposed to. Over a decade later, when writer-producer J. Michael Straczynski was formulating his classic sci-fi 'novel for TV', *Babylon 5*, one of his earliest key edicts was 'no cute kids, no cute robots'. One can't help but think he must have had Twiki as an unpleasant memory.

HECTOR in *Saturn 3* (1980)
Not exactly the highlight of actor Kirk Douglas's career, as he is kitted out in a tinfoil suit and forced to do battle against a rubbish 8ft robot in deep space. Hector has unfortunately tapped into the brain of the crazy Captain Benson, and ends up doing all sorts of silly and un-robot-like things.

GREEN CROSS CODE ROBOT (early 1980s)
Of course we should support the worthy message behind this one, but streetwise superhero and lycra-clad West-Country road safety expert The Green Cross Code Man always looked a little uneasy accompanied by his automated sidekick. The Green Cross Robot appeared to have been rather shoddily designed from a cartoon template of what Artoo Detoo might have looked like if he was channelling Metal Mickey, and trundled across the road in a somewhat ungainly fashion. No surprise, then, that he was replaced in short order by a Grandmaster Flash parody ('Don't Step Out When You're Close To The Edge'…) featuring some hideously vibrant use of electronic Paintbox.

V.I.C.I. in *Small Wonder* (1985-89).
Did we ever really buy the rather creepy premise of this Metromedia/20th Century Fox production – that the robotic Victoria 'Vicki' Ann Smith-Lawson is modelled on a real girl and lives with an ordinary family? On similar lines is D.A.R.Y.L., from the 1985 film of the same name, perhaps showing us that cute/creepy kid robots are really not the way to go.

ANDREW/NDR in the film *The Bicentennial Man* (1999), based on the novel by Isaac Asimov.
Played by Robin Williams in a metallic robot suit. Harsher critics have claimed that the script suffers from similar constraints. The story of the voyage of discovery by a robot who is adopted into the Martin family to perform household duties, and who

starts to find he has outlived his usefulness. Although the *LA Times* called the film 'sumptuous' and praised its design and effects, veteran critic Roger Ebert found that it 'finally sinks into a cornball drone of greeting-card sentiment', while the *Philadelphia Inquirer* thought it 'excruciating, ponderous, remarkably unfunny and inert.'

THE *EXTERMINATOR CITY* ROBOTS (2005)

In this low-budget film from the decade of snazzy effects and CGI, all the robots are either people in costume or puppets. Right. The nadir is reached when two robots – not even named in the credits – square up to one another. Scary red eyes glowing, and with a bit of growled dialogue, they engage in what must be the wimpiest swordfight ever seen in the cinema. Errol Flynn would be rolling in his grave. It's supposed to look fast and furious, as if these automata have lightning-fast responses – instead it simply looks (and sounds) as if they are clinking cutlery together prior to carving up a bit of gammon. It's truly embarrassing.

And now, back to some more mechanised men, women and beasties who were memorable for all the *right* reasons…

Chapter 5

The Age of Paranoia

MR. FREEZE

Also known as: Mr Zero.

Key Narratives: *Batman* in DC Comics from 1958 as Mr Zero, from 1968 as Mr Freeze, then *Batman* TV series and movie franchise (*Batman and Robin*, 1997).

Appearance: Blue humanoid in cryogenic suit with tight-fitting helmet.

Design Classic? Has been updated over the years, from his initial appearance in a glass helmet and greenish suit to the icy-gothic costume of the character as played by Arnold Schwarzenegger in the films.

Origins: A scientist whose experiments with an 'ice gun' backfire and end up with his being doused in cryogenic chemicals and needing to be kept at sub-zero to survive. Later reinvented as a molecular biologist who enjoys freezing small animals.

Designation: Human cyborg.

Weaponry: Ice Gun – you can guess what it does. Refrigeration suit granting him protection, strength and near-immortality.

Fear Factor: 👽 👽 (Originally intended to be a bit of a 'joke villain', but made more fearsome in later incarnations.)

Cuteness Factor: ♥

Artificial Intelligence: High – another of those brainy science bods gone bad.

Skills: Messing about with molecular structures and doing sub-zero stuff.

Catchphrase: 'Revenge is a dish best served… cold.' The origin of this expression can be traced back to one associated with (although which does not appear in) the eighteen-century epistolary novel *Les Liaisons Dangereuses* by Pierre Choderlos de Laclos (filmed in 1988 as *Dangerous Liaisons*).

Notability: ★★

Databank
- His real name is Dr. Victor Fries, and he was invented by comics writer and artist, and *Batman* creator, Bob Kane for the *Batman* comic in 1958.
- He was played by various actors in the TV show; George Sanders, Otto Preminger (who is very fond of the word 'Wild!' in his performance) and Eli Wallach. Michael Ansara voiced him for the animated series, and in the 1997 film *Batman and Robin* he was played by Arnold Schwarzenegger.
- Mr Freeze is technically a cyborg as the technology which keeps him alive is essentially part of him – the suit is integral to his survival. (It's arguable, but…)
- The *Blackhawk* comic strip 'The Fantastic Mr Freeze' features a 'Robot Mr Freeze', controlled by one Professor Thurman – a different character and not to be confused with this villain.

Further Appearances
Various video games and other spin-offs.

Assessment
Another larger-than-life comic-strip villain whose history has been through so many re-inventions and permutations that it could be argued there is no 'definitive' version. Sure to make a re-appearance in the franchise at some point, although his method of causing mayhem is something of a giveaway…

Verdict: Ice, ice, baby.

ALICIA

Key Narratives: 'The Lonely', from the TV anthology series *The Twilight Zone*.

Appearance: Human female.

Origins: Sent by Earth to the anti-hero, criminal Corry, who is in solitary confinement on a lonely asteroid outpost for 50 years.

Designation: Android, non-hostile, companion.

Weaponry: n/a

Fear Factor: ♥

Cuteness Factor: ♥♥♥ (She is a young Jean Marsh, after all…)

Artificial Intelligence: High.

Skills: Making conversation. Possibly also making tea. Yes, all right. It's set in 2046 but this was the pre-feminist 1960s.

Notability: ✯✯

Databank
- One of the earliest entries in the *Twilight Zone* canon. The 1960s series was an anthology of half-hour short stories each with some sci-fi, fantasy or paranormal twist, presented (and usually written) by Rod Serling.
- In this story, the isolated criminal James Corry, played by Jack Warden, is visited four times a year by a spaceship captained by Allenby (John Dehner) which brings him supplies and news from Earth. One day a special delivery comes in the form of robot 'companion' Alicia, played by Jean Marsh – a kindness by Allenby to help alleviate Corry's sense of loneliness.
- The twist comes when Corry is pardoned, and can return to Earth, but finds his luggage allowance doesn't include Alicia. He tries, unsuccessfully, to argue that she is human. Allenby, most uncharitably, shoots Alicia, exposing her wires and circuitry… Thankfully this doesn't have the effect of sending Corry into a homicidal rage, but reminds him that all he will be leaving behind is his loneliness.

Quotes
'Witness if you will a dungeon, made out of mountains, salt flats and sand that stretch to infinity. The dungeon has an inmate: James A. Corry. And this is his residence: a metal shack. An old touring car that squats in the sun and goes nowhere, for there is nowhere to go. For the record, let it be known that James A. Corry is a convicted criminal placed in solitary confinement. Confinement in this stretches as far as the eye can see, because this particular dungeon is on an asteroid nine million miles from the Earth. Now witness if you will a man's mind and body shrivelling in the sun…a man dying of loneliness.' – Opening narration by Rod Serling.

Assessment
One of the earliest *Twilight Zone* episodes, and a very well-regarded one too. Like many of the entries in the anthology, it explores the borders of madness and sanity, the impact of human isolation and, ultimately, like many stories of people and robots, what it actually means to be human.

Verdict: Abandoned.

ROBERT

Also known as: Robert the co-pilot.

Key Narratives: Space puppet 'supermarionation' series *Fireball XL-5*, first shown in 1962.

B: See-through robot with a spindly body and clamp-like hands. Head shaped like an upturned bucket with an antenna.

Design Classic? Not really – the transparent look never really caught on…

Origins: Earth's most advanced robot, invented by Professor Matthew 'Matt' Matic. He is made from a substance called 'leadinium', as we are told in the episode 'Robert to the Rescue'.

Designation: Co-pilot.

Weaponry: Hand-held gun.

Fear Factor: 👽

Cuteness Factor: ♥♥

Skills: Piloting.

Notability: ✯✯

Databank
- Robert was voiced by *Fireball XL5* creator Gerry Anderson himself. The spaceship *Fireball XL5* is part of the World Space Patrol and is based in a scientific establishment called Space City. It patrols Sector 25 of space and visits a number of different planets in the series. Some dialogue indicates that Space City is in the Pacific Ocean.
- *Fireball XL5* briefly had the title of *Nova X 100* before Gerry Anderson renamed it. He claimed that his inspiration for the eventual title came from the motor oil Castrol XL.
- The series was all shot in black-and-white, but one episode, 'A Day in the Life of a Space General', was colourised as an experiment for the DVD release.

Assessment
Robert's quirky appearance has definitely inspired his fans. There are various websites devoted to making replicas of him. At the time, he would have seemed both very futuristic and retro – a homage to old-fashioned sci-fi.

Verdict: Bobby dazzler.

DOCTOR OCTOPUS

Also known as: Dr Otto Gunther Octavius; 'Doc Ock'.

Key Narratives: The *Spiderman/Spider-Man* comics and films.

Appearance: Stockily-built, short-sighted and with several mechanical appendages.

Design Classic? A classic of comic art, which has certainly inspired the movie designers to build on it and have some fun with it.

Origins: First appeared in *Amazing Spider-Man* comic in 1963. Brought up by an over-protective mother and a violent father (why do all American villains have to have Daddy Issues?), young Otto was shy and hard-working in school, becoming a top student. He became a respected scientist and developed a set of mechanical arms with a brain-computer interface. But in the weird world of Marvel Comics, bodily transformation is never far away, and indeed after a radiation accident, the mechanical arms become fused with his physical form. The accident re-wires his brain and he takes to a life of crime.

Designation: Highly intelligent, part-mechanised, mad scientist supervillain.

Weaponry: His set of deadly mechanical extrusions – four powerful, near-indestructible tentacles – which he controls with the power of his mind.

Fear Factor: 👽 👽 👽

Cuteness Factor: ♥

Artificial Intelligence: High – genius level.

Skills: Inventor, atomic physics researcher, super-criminal.

Notability: ★★★

Databank

- Dr Octopus is near-sighted, but this doesn't impede his ability to battle our hero; in fights he is a match for Spiderman.
- It's implied first of all that the telepathic control over his limbs is a result of his mind re-adjusting itself after the accident. Later stories hint that he may have possessed low-level telepathy from his youth, which was in turn implied to have caused a brain aneurysm in his father.
- He briefly abandons his Doctor Octopus identity and creates the identity of the 'Master Planner', a criminal mastermind, to orchestrate his plots.
- Along with five other super-criminals – The Vulture, Kraven the Hunter, Electro, Mysterio and The Sandman – Dr Octopus forms the Sinister Six, first introduced in the 1964 annual. The Sinister Six has had several line-ups over the years, even becoming a Sinister Seven at one point, and a sadly non-alliterative Sinister Twelve at another.
- Dr Serena Patel is the female version of Dr Octopus who appears in the Marvel 2099 possible-future timeline. She is voiced in the video game by Tara Strong. Other versions of him include Baron Octavius, an Italian nobleman from the Marvel 1602 universe.

Further Appearances
Appears in the 1960s TV series, voiced by Vernon Chapman, and in the 1990s TV series voiced by Efrem Zimbalist, Jr. Played by Alfred Molina in the 2004 film *Spider-Man 2*. Also appears in numerous video-game and computer-game spinoffs.

Nagging Questions
Just how does the 're-writing' of thr Doc's brain actually happen? It's thought at first that the radiation accident has sent him mad, but it's just his mind re-adjusting to having four large mechanical extrusions 'fused' into him… a pretty large re-adjustment, we'd have thought.

Memorable Moment
Peter Parker, aka Spiderman, is shaken by his first encounter with Octopus, in which the tentacled supervillain swats our arachnophiliac hero aside (with his human hand), sending him tumbling out of a window. It's the first sign that he is a villain to be reckoned with… and Spiderman's first real moment of self-doubt.

Assessment
As with all Marvel supervillains, one is required just to suspend disbelief and go with the flow, abandoning any pretence at scientific believability at the door. Comic-book narrative is hugely enjoyable if one accepts its conventions. The trouble with all superheroes is that they need worthy opponents who are almost, but not quite, their match, and in this respect, the Doc steps up most admirably. And as with all the best villains, he is given an extra dimension by the undercurrent of his tragic past, and the occasional resurgence of a more human side to his character.

Verdict: Criminal intent-acle.

THE DALEKS

Key Narrative: The *Doctor Who* TV series, books and films. First properly appeared in the fifth episode 'The Survivors' in 1963, and have continued to appear in the series intermittently up to the present day. Every Doctor apart from the eigth (Paul McGann) has faced the Daleks on screen at least once, and they always seem to bounce back despite being defeated.

Appearance: Squat, pepper-pot shapes approximately 1m 60cm high with a domed head from which an eye-stalk protrudes. The torso is mounted with a gun-stick and a telescopic arm with a sucker-like attachment. The Dalek moves by gliding along and can also float ('elevate') at some speed to overcome obstacles such as stairs and lift-shafts.

Design classic? Absolutely. Survived several minor re-designs down the years.

Origins: Several versions of Dalek history exist. However, the commonly accepted one was established in the 1975 story 'Genesis of the Daleks' written by Terry Nation and directed by David Maloney, in which the Daleks are developed on the planet Skaro by the brilliant crippled scientist, Davros. Foreseeing what the Kaled race would become, Davros conducted a number of hideous experiments in his Bunker and developed the Dalek casing as a 'Mark 3 Travel Machine' casing to house the Kaled mutant creature inside. The Daleks turned on Davros, but later required his help and co-operation on several occasions (Destiny of the Daleks in 1979, 'Resurrection of the Daleks' in 1984 and 'The Stolen Earth/Journey's End' in 2008). An uneasy relationship exists between Davros and his creations.

Designation: Cyborgs – emotionless, ruthless and logical.

Weaponry: Gun-stick, capable of dispensing deadly laser bolts. No external scarring is seen to the Dalek's victims, and it is established in 'Rememberance of the Daleks' (1988) that the Dalek ray causes death by massive disruption to internal organs. In 1974's 'Death to the Daleks', the Dalek weapons were incapacitated on the planet Exxilon by the power-drain from the City of the Exxilons. However, they re-asserted their superiority by re-arming themselves with simple percussion weapons resembling built-in machine-pistols. In the novel *Prisoner of the Daleks*, it's established that each Dalek could reduce a human to atoms in a millisecond with its gun, but instead, chooses a lower setting which makes the human's death last a couple of seconds in order for it to be all the more agonising.

Fear Factor: ☻ ☻ ☻ ☻ ☻ (Look, how high do you want to go?!)

Cuteness Factor: ♥ (Some people still love them…)

Intelligence: Extremely High – devious and cunning.

Moral guidance: Emotionless, driven by desire to become the supreme race in the Universe and/or annihilate all other races.

Catchphrases: 'Exterminate!' and 'You will obey!'

Notability: ★★★★★

Databank

- The Daleks, like their *Doctor Who* stablemates the Cybermen, are technically cyborgs, as they are made up of organic and robotic components. Little is known about the way in which the Kaled mutant inside the shell interacts with the robotic components.
- In 1979's 'Destiny of the Daleks', it is stated that the Daleks have evolved into robots. Given that this story is scripted by their original creator, Terry Nation, one must assume that this is intentional and not a slip-up, and that this stage of

evolution remains at least a possibility for future Daleks (even if the twenty-first-century brand are most definitely still part-organic).
- The Daleks' grating voices and their battle-cry of 'Exterminate!' have become known to generations of TV viewers.
- BBC designer Ray Cusick came up with the look of the Daleks, based on minimal descriptions in Terry Nation's scripts.
- The Dalek 'operator' is an actor who sits inside the casing, moves it around, is able to see very little, and is reliant on the director's instructions. The Dalek voices are done by a separate actor – variously, over the years, Roy Skelton, Royce Mills and Nick Briggs – equipped with a device called a Ring Modulator to produce the classic harsh, grating tones.
- From 2005 onwards, it is established that the organic matter inside the Daleks has been 'harvested' from rejected, mutated and disabled members of the human race.
- The Daleks and the Cybermen did not confront one another on screen until the 2006 episode Doomsday.
- Since 2010's 'Victory of the Daleks', a re-designed form of the Daleks has been seen. These are bigger and chunkier, more tank-like, although they still have the basic pepper-pot shape. Each type of Dalek sports different coloured livery and their grating voices are now deeper and more resonant.
- The Dalek system of justice has never been fully explained, but on at least two occasions they have taken an individual back to their home planet for 'trial'. One was their creator Davros, and the other, the Doctor's arch-enemy The Master.
- A standing joke among lazy journalists in the 1970s and 1980s was the Daleks' supposed helplessness when faced with a staircase. However, the observant viewer will note that the Daleks climb stairs by implication in the 1965 story 'The Chase', have hover-pads on which they can float by 1973's 'Planet of the Daleks', and are first seen to propel themselves up stairs with internal 'hover-jet' systems in 1988's 'Remembrance of the Daleks'. By the time of 2005's 'Dalek', floating and flying is second nature to them. It's doubtful that the new generation of Who fans would even get the joke.
- *Who* fandom is mixed in its reception of the new, chunkier, multi-hued Daleks. Detractors claim that they have lost their elegant sleekness, with the addition of a 'hump' on the back, and that the bright colours make them look like the Teletubbies or some kind of 'Daleks Aloud' colour-co-ordinated band. Supporters point out that the Daleks have always subtly altered their look over the decades, that the essential shape remains the same, and that the chunkier, more tank-like look of the new models, together with their booming and resonant voices, gives them more screen presence. Time will tell.

Further Appearances
- In addition to the TV series, two *Doctor Who* films were made in the 1960s, loosely based on two of the William Hartnell stories and starring Peter Cushing in the role of the Doctor. These were *Doctor Who and the Daleks* (1965) and *Daleks: Invasion Earth 2150 A.D.* (1966).
- Several official *Doctor Who* novels and CD audio dramas have featured the Daleks, although copyright issues with the estate of Terry Nation have prevented their being used as often as writers in these media might have liked.

Nagging Questions
The continuity of Dalek history has always been a thorny question, thanks to its being cobbled together from stories written by several different writers. The story *The Evil of the Daleks* (1967) was supposedto depict the final end of the Doctor's greatest enemies, but now that it is clearly established that they, too have the power of time-travel, anything is possible.

Their invasion of Earth in the twenty-second century may, now, never have happened in Doctor Who history, given the events of Day of the Daleks (1972) centring around a group of time-travelling guerrillas and their attempts to change history. As with anything involving time-travel, it all gets very confusing very quickly.

It is likely that we will never find out what went on during the Time War, the great conflict between Daleks and Time Lords in which many races were involved, and in which the Doctor's actualy role remains vague…

Memorable Moment
The Doctor runs up the cellar stairs and the Dalek, propelled by some kind of energy underneath it, rises for the first time to follow him. A thousand journalists' cheap jokes evaporate in one sublime moment of television.

Assessment
Immune these days, from the lazy jokes about sink-plungers, egg-whisks and the dangers of staircases, the Daleks of the twenty-first century remain as threatening as ever to children of all ages. They are TV's most iconic cyborgs, recognisable even to sci-fi haters, and as iconically representative of the BBC as Broadcasting House itself. It's amazing that, over the course of five decades, the Daleks have stood the test of time and bounced back again and again, surviving explosion, fire, plague, war, ridicule and the programme's cancellation (twice).

Verdict: Love to hate them.

'ROBOT' (from *Lost in Space*)

Also known as: B-9, Class M-3 General Utility Non-Theorizing Environmental Control Robot. Unsurprisingly, this somewhat wordy name is mostly abandoned in favour of just 'Robot'. Shame they couldn't come up with a handy acronym, but there you are.

Key Narratives: Irwin Allen's *Lost in Space* TV series 1965-68, famous for its space-age action and wobbly flying saucers

Appearance: Similar to Robby from *Forbidden Planet* – no coincidence, as they share a designer, Robert Kinoshita. 'Robot' was approximately 6ft high, with concertina-like limbs, a solid cylindrical chest/trunk unit and a swivelling transparent head.

Design classic? Not so much as Robby. A bit unwieldy.

Designation: Helpful.

Weaponry: An array of weapons including laser beams and a crackling 'electro-force'.

Fear Factor: 👽 👽

Cuteness Factor: ♥♥♥ (Robot is more 'awww' than 'awe'.)

Artificial Intelligence: Very high. Incredible recall of facts, analysis skills, calculations, deductions, etc.

Skills: As well as the above, could mimic human traits such as laughter and musical ability (e.g. playing the guitar and operatic singing).

Catchphrase: The most famous of all, perhaps: 'Danger, Will Robinson! Danger!' So well-known it is often quoted by people who have never even seen the programme. And, like 'Beam me up, Scotty', it's never actually spoken in the show in that exact format. 'Danger, danger!' was often said by Robot, and 'Danger, Will Robinson!' just once. He also says 'Warning, warning!' quite a bit.

Notability: ★★

Databank
- Three men were instrumental in bringing Robot to life – designer Robert Kinoshita, actor Bob May inside the suit, and voice artist Dick Tufeld.
- Dick Tufeld's career began as a radio voice artist and announcer, and he worked on the famous 1950s TV series *Zorro*. He died in January 2012.
- There were actually two Robot costumes, one of which was kept for stunts only.
- The light-hearted space adventure programme told the story of the Robinson family who set off from an over-populated Earth in 1997 (cunningly escaping New

Labour and the Elton John tribute to Princess Diana in the process) in search of a new home. The first series was in black-and-white, the rest in colour.
- The Robot is sabotaged by Professor Zachary Smith, a character who was going to be killed off once defeated but is kept on afterwards largely for comic relief.
- The theme music was provided by the famous John Williams, who went on to provide the score for *Star Wars*, *Indiana Jones*, *Superman* and many more.
- The aforementioned Robby – who popped up all over the place after *Forbidden Planet* – guest-starred in two episodes, 'The War of the Robots' and 'The Condemned of Space'.
- *Lost in Space* was quite an expensive show for the time (the pilot episode cost $600,000, which for a pilot was a budget only beaten then by *Star Trek*'s 'The Cage', which clocked up $630,000).

Quotes
- The villainous Dr Zachary Smith had a selection of alliterative insults which he'd lob in Robot's direction, including 'mental midget', 'ludicrous lump' and 'bubble-headed booby'.
- 'My micromechanism thanks you, my computer tapes thank you, and I thank you.'

Further Appearances
- Robot has a website dedicated to him at www.lostinspacerobot.com where you can get hold of a full-size, limited-edition Robot replica, should you be willing to do so…
- http://www.b9robotresource.com/ also has a guide to building a replica model.
- ABC produced a *Lost in Space* cartoon in 1972 as part of a series of animations called *The ABC Saturday Superstar Movie*. It featured Dr Smith and a version of Robot called Robon.
- The 1998 *Lost in Space* film was quite successful, with Dick Tufeld again providing the voice of Robot. By this stage Robot has evolved into a larger, more hulking automaton with broad shoulders, like some kind of silvery American Footballer, with large, gripping, claw-like hands.
- In 2003 a pilot for a revamped, remade TV series was filmed, and featured a new, sleeker, more anthropomorphic Robot. It was not picked up for a series and remains unaired.

Assessment
The Robot, sticking with the original, embodies the 1960s vision of what the future would look like, and, like Robby, is a homage to B-movie and Republic serials' larger-than-life space-adventure shenanigans. The science by which he works is accepted and not explored in any great depth – it is simply accepted that this advanced piece of walking, talking technology has these wondrous abilities. Perhaps in 1965 it was thought that we'd all have these in our homes by the late twentieth or early twenty-

first centuries. It's rare for dated sci-fi to have predicted advances such as the internet, wi-fi and powerful data-processing. However, an abundance of robots with almost-unlimited data retrieval does hint that some writers were allowing their imagination to go down that route.

Verdict: A useful companion in space…

THE CYBERMEN

Also known as: Mondasians, Telosians

Key Narratives: *Doctor Who*, notably the TV stories 'The Tenth Planet' (1966), 'The Tomb of the Cyberman' (1967), 'The Invasion (1968), 'Earthshock' (1982), 'Rise of the Cybermen/The Age of Steel' (2006), and others, plus numerous novels and audio and comic-strip spin-offs. By various writers including Gerry Davis and Kit Pedler, Eric Saward, Russel T. Davies, Gareth Roberts.

Appearance: Silvery, technologically-enhanced humanoids, tall, with blank faces and jug-handled ears, plus a 'chest unit' containing complex workings.

Design classic? Design changes considerably over the years, but generally holds up well.

Origins: In the original *Doctor Who*, shown first to be from the Earth's twin planet Mondas, then from the planet Telos where their ice tombs are defrosted. In the new twenty-first-century series they are created in a parallel universe by John Lumic.

Designation: Cyborgs: relentless, emotionless.

Weaponry: Extraordinary-strength hands capable of crushing human flesh and bone. Various hand-held weapons shown over the years, including slim, rod-like guns in the 1960s and chunky blaster rifles in the 1980s. 'Revenge of the Cybermen' (1975) shows them blasting deadly rays from their head-pieces.

Fear Factor: 👽👽👽👽👽 (They could climb stairs, very fast, long before we ever knew the Daleks could…)

Cuteness Factor: ♥ (If only for the 'crying' moment in 'Doomsday'…)

Artificial Intelligence: High.

Skills: Logical assessment, battle and combat.

Catchphrases: 'You will be like us!' and later 'Delete!'

Notability: ★★★★

Databank

- First seen lumbering out of the swirling blizzard in William Hartnell's last Doctor Who story, 'The Tenth Planet' the Cybermen have survived almost five decades to become as iconic and as much a staple of children's nightmares as the Daleks.
- The science behind the Cybermen was intended to be realistic. Their creators were writer Gerry Davis and medical scientist Kit Pedler, the show's unofficial 'scientific adviser'. Pedler was fascinated by the science of cybernetics and by how much of our humanity we would retain if we replaced our unreliable organic components, one by one, with artificially-created ones. The Cybermen were the logical, nightmare extension of the cybernetics – coldly calculating, with no human emotion left at all.
- The original Cyberman designs very much stress their humanoid origins, with a silvery-grey cloth mask stretched across the face, surrounded by a chunky head-unit sporting a large lamp-like attachment. Their hands look fully organic – as far as one can tell from black-and-white recordings.
- Subsequent re-designs took them further and further away from the human silhouette; the head becoming squarer and the body more streamlined. The addition of a 'teardrop' shape on the eyes was a clever touch; a reminder of the emotions the Cybermen had lost. By the 1980s, with their ungainly gauntletted hands and sleek glass jaws, the Cybermen looked more robotic than ever. Their twenty-first-century reinvention takes their form back, in some respects, to their original 1960s look, but with a chunkier, threateningly bulky 'American footballer' build.
- The Doctor finds various ways of defeating the Cybermen over the years. The original bunch are despatched with radiation, while the squad which tries to invade the Moon are launched off into space by means of the weather-controlling Gravitron machine. They are impervious to bullets, but some succumb to heavier artillery in *The Invasion*. By the 1970s, it has been established that the Cybermen have an aversion to gold – taken to extremes in 1988's 'Silver Nemesis' in which teenage assistant Ace manages to dispose of several Cybermen with gold nuggets fired from her catapult.
- The Cybermen are sometimes assisted by the silvery, rodent-like Cybermats, which first appear in two 1960s stories, 'The Tomb of the Cybermen' and The Wheel in Space' and return in 1975's 'Revenge of the Cybermen' and 2011's 'Closing Time'.
- In its first series, *Doctor Who* spin-off *Torchwood* somewhat implausibly – and some say laughably – presents us with the notion that Torchwood team regular Ianto Jones has been looking after his half-Cyber-converted girlfriend Lisa Hallett in the basement of their HQ. Predictably, all hell breaks loose, and a bloody and violent confrontation results in the death of the 'Cyberwoman'.

- The Cybermen are almost certainly one of the inspirations for *Star Trek*'s Borg. Ironically, with *Doctor Who* having been off the air for a lot of the period of Trek's greatest success and then returning, some Trekkies tried to put about that the reverse was true…

Quotes
'We must survive! We must survive!' – The Cyber-Controller.
'Delete! Delete!' – Cyber-catchphrase since 2006.

Further Appearances
Official audio plays, notably *Spare Parts* (2002), and various official novels including *Iceberg* (1993) and *Made of Steel* (2007).

Nagging Questions
What exactly does the Cyber-conversion process involve? Originally the Cybermen were said to have 'replaced' their organs gradually, and now it is implied in some stories that the human brain is extracted and placed inside the Cyber-casing, as if it were a futuristic suit or armour. The attempted conversion of Craig in 'Closing Time', foiled by his love for his son, shows his entire body being encased in the Cyber-suit.

Fans have debated for years as to whether Mondas or Telos was the Cybermen's original planet, with evidence being produced on both sides. The 'John Lumic' Cybus generation of Cybermen, produced in an alternative universe and only able to access ours by means of a dimensional wormhole, add another layer of complication. Their origins in the parallel universe have, it's fair to say, been largely ignored by more recent episodes.

Memorable Moment
Their dramatic, cliff-hanger return at the end of Part One of Peter Davison's *Earthshock* (1982). Genuinely unexpected, and a key memory for fans who were watching at the time.

Assessment
Despite their origins in 'real science', the need for the Cybermen to be convincingly hostile adversaries has, over the years, somewhat compromised the purity of creators Davis and Pedler's intentions. Their outings in the 1980s moved the Cyber-threat away from the body-horror aspect and made them more generic 'recurring baddies'. While the new episodes have tried to concentrate on what makes the Cybermen 'different', they too have occasionally been guilty of shoving them into stories where they don't really do anything interesting.

Verdict: Still menacing, but is their finest hour yet to come?

THE CYBERNAUTS

Key Narratives: TV series *The Avengers* (1961-69) and *The New Avengers* (1976-77), specifically the episodes 'The Cybernauts', 'Return of the Cybernauts' and 'The Last of the Cybernauts…??'

Appearance: Blank-faced, black-hatted, black-coated, shades-wearing and humanoid in shape, like robotic spies.

Design classic? Suitably menacing, but perhaps a rather simple look.

Origins: Invented by a chillingly mad scientist called Dr Armstrong, played by Michael Gough.

Designation: Hostile robots.

Weaponry/Skills: Strength and combat. In *Return of the Cybernauts* the new Cybernauts are directed by heartbeat patterns.

Fear Factor: 👽 👽 👽

Cuteness Factor: None, really. Unless you like their fashion sense.

Artificial Intelligence: n/a

Notability: ★★★

Databank

- Long before the word was popularised as a (now slightly archaic) term for a keen internet-user, a Cybernaut was something else entirely. First glimpsed smashing their way into an office to murder an electronics executive, the Cybernauts embody the traditional *Avengers* values of adventure, menace, camp and over-the-top fun. Despite their recurrence, they don't quite become an 'arch enemy' for the Avengers in the way the Daleks and Cybermen do in *Doctor Who*.
- The first Cybernauts story, interestingly, makes a number of predictions about technology which have since come true (the character Tusamo speaks of 'Computers no bigger than a cigarette box. Pocket television. And radios smaller than a wristwatch').
- When they return in *The New Avengers*, the Cybernauts have been re-activated by evil double-agent Kane, who is looking for a means to inhabit a robot body.
- The 1975 *Flash Gordon* novel *War of the Cybernauts* is nothing to do with this lot.

Quotes

Emma Peel: *'It's a karate blow. Delivered by an expert, it breaks the neck easier than a hangman's noose.'*

Nagging Questions
The ending presents us with Cybernaut with a brain of its own. It seems odd to introduce this right at the end, almost as a casual add-on. What implications will this have for future Cybernauts?

Assessment
Lean, mean killing machines of steel, perhaps not as scary as the Terminator or the Cybermen, and not as iconic, but memorable for all the right reasons. It's surprising they didn't appear in more *Avengers* episodes.

Verdict: Not so very Cyber, but still chilling.

THE IRON MAN

Also known as: The Iron Giant.

Key Narratives: The 1968 book *The Iron Man* by Ted Hughes, and the 1999 film version. This was named *The Iron Giant*, as was the US edition of the book, so as to avoid confusion with the Marvel Comics character Ironman.

Appearance: A giant metal man who is 'taller than a house', with a head 'shaped like a dustbin' and 'as big as a bedroom', with 'eyes like headlamps'.

Origins: 'Nobody knows', according to the story. He has never seen the sea before, which suggests he comes from a waterless world.

Designation: Giant robot.

Weaponry: n/a

Fear Factor: 👽 👽

Cuteness Factor: ♥♥

Skills: Great strength and, indirectly, the bringing of peace and harmony.

Notability: ✹✹

Databank
- The Iron Man arrives seemingly out of nowhere. To survive, he feasts on farm machinery, and he promises not to bother the local populace if they don't bother him.
- When a creature appears in the skies (a 'Space-Bat-Angel-Dragon'), the Iron Man, who has been accepted as part of the community, challenges the creature to a trial of strength involving surviving great heat. After it is defeated the creature is revealed as a 'Star Spirit' which brings harmony to the universe through the Music of the Spheres.

- The film version boosts the tension a bit, locating the story in the USA at the height of the Cold War and making the Iron Man's initial attacks on machinery more dramatic. A boy called Hogarth Hughes realises the robot is self-repairing and teaches it all about heroism, but the US Army are on their trail… The story ends with the Iron Giant's self-sacrifice.
- Pete Townshend of The Who produced a 1989 rock musical based on the book, which indirectly led to the film.
- The book was also read on the BBC's *Jackanory* by none other than Doctor Who himself, the legendary Tom Baker.
- *The Iron Man* is often used as a key text in schools and features on the BBC Learning Zone.
- Not to be confused with the other 'Iron Man', aka Tony Stark, the Marvel Comics superhero, who is an injured engineer living inside a powered suit of armour packed with crime-fighting weapons. He originally appeared in the *Tales of Suspense* comic in 1963.

Quotes

Despite the major alterations from his book, Hughes was pleased with the film script, saying in a letter: *'I want to tell you how much I like what [director] Brad Bird has done. He's made something all of a piece, with terrific sinister gathering momentum and the ending came to me as a glorious piece of amazement. He's made a terrific dramatic situation out of the way he's developed The Iron Giant. I can't stop thinking about it.'*

'There's plenty of humor, and the whole movie is laced with intelligent asides… When push comes to shove, the big iron guy reveals a built-in capacity for destruction, and the plot turns smartly on a showdown that easily could have become another formula for cartoon mayhem.' – San Francisco Chronicle reviewing the 1999 film

Further Appearances

In 1993 Hughes wrote a sequel, *The Iron Woman* (*it's* not too difficult to guess the nature of the protagonist). The sequel has more of an environmental subtext than the first book. The heroine Lucy befriends a giant iron woman who emerges from a local swamp covered with poisonous chemicals. The Iron Woman is on a mission to stop world pollution by destroying the factories which harm the natural world.

Memorable Moment

The Iron Man makes quite an epic, thunderous entrance in the original story, appearing at the top of a cliff, which he walks over the edge of. He tumbles downwards, falling to pieces, and lies in bits on the beach. The bits began to locate one another…

Assessment

As science-fiction the story is fairly simple, although it can also be seen as a simple moral fable for the entertainment of children.

Verdict: Tough as nails, but with a softer side.

Chapter 6

Double Trouble

The Best Ever Robot Doppelgängers

Whether they call themselves duplicates, doubles, replicas or clones, robot impersonators are the ultimate in robot-technology. A robot not just called to do the boring jobs humans don't like, or go into battle with fast reflexes, but rather to take the place of an organic life-form and assume its role – usually for the purposes of deception.

Superman robots: in the 1950s and 1960s comics.
Unusual in that they were not created by an evil supervillain, but rather to deputise for Superman himself on some missions. One would think the Man of Steel, who after all has many of the abilities of a robot himself (super strength, combat skills, X-ray vision, etc.) wouldn't need to send robot ringers to do his work, but then again he does have that irritating susceptibility to Kryptonite, which you'd really think he'd have got sorted out by now. Also used for protecting his identity. The robots didn't possess all of his powers. Perhaps he used to send some of them to do the boring stuff, like shopping. And tax. And those long chats with Lois Lane about her feelings.

Android Kirk: from the original *Star Trek* (episode 'What Are Little Girls Made Of?' from 1968).
It seems Nurse Chapel, often seen bustling around the *Enterprise* Sickbay but rarely allowed to have any actual character development, had a secret past – a man called Roger. Being an enterprising chap, Roger made the most of finding an alien civilisation and used their technology to build himself an android body. But he didn't stop there – no. Roger thought it might be a good wheeze to construct an android duplicate of everyone's favourite emoting, snogging and baddie-thumping Captain, James T. Kirk. The double wasn't terribly convincing. Although it was physically exact, it came out with things like, 'Mind your own business, Mr. Spock! I'm sick of your half-breed interference, do you hear?' – which alerted the ever-vigilant half-Vulcan First Officer to the fact that something wasn't quite right. The fake Kirk was destroyed by the lovely Andrea, also a robot.

Mechagodzilla: from the film *Godzilla vs. Mechagodzilla* (1974).
You don't even have to be human(oid) to suffer the curse of the robot double. Roaring, city-destroying, nuclear-powered, ravenous reptile of doom Godzilla encountered his own fearsome facsimile in 1974, thanks to one of those incredible team-ups which movie directors love so much. Mechagodzilla was created by ape-like lifeforms from the Third Planet From the Black Hole (yes, it was), in order to defeat Godzilla and conquer Earth. Godzilla eventually defeats the iron impostor by means of drawing it closer in a magnetic field, and then snapping its head off. Like some giant Japanese reptilian Ozzy Osbourne.

The android Sarah Jane Smith: from *Doctor Who* 'The Android Invasion' (1975).
Perhaps the most memorable of the many doubles brought to us by everyone's favourite teatime sci-fi show. The 1970s stories loved to pastiche the tropes of horror movies, and this particular script by Terry Nation had Tom Baker's Doctor and his companion Sarah-Jane Smith (Elisabeth Sladen) stumbling across an android version of *Invasion of the Body-Snatchers,* controlled by a race of pig-like aliens called the Kraals. Halfway through the story, the Doctor, who's no fool, guesses that the 'Sarah' who has conveniently escaped the aliens' clutches is a fake, as is the entire village and the countryside around it. He snaps a twig in two to show it's plastic and tells Sarah she's not the real thing – at which point she pulls a gun on him. In the ensuing struggle, her plastic face falls off, revealing the circuitry beneath – a visual image imprinted on the minds of seven-year-olds everywhere!

The Buck Rogers clone: from the *Buck Rogers in the 25th Century* episode 'Ardala Returns' (1980).
The evil Ardala, aided and abetted by the Draconians, creates a near-perfect clone of Buck. However, like the Kirk clone created by Roger, it isn't quite right. (You'd think these people would know how not to slip up, wouldn't you?) The robot mangles his twenty-first-century slang and doesn't quite get his personality right. It's sent to Earth rigged with a bomb, but good old Colonel Wilma Deering senses something is up and disposes of the fake Buck pretty sharpish.

The robot Bill & Ted: from *Bill and Ted's Bogus Journey* (1991).
Hapless time-travellers Bill (Alex Winter) and Ted (Keanu Reeves) face their evil doubles, created by villainous Joss Ackland to prevent them from winning a Battle of the Bands. The robots don't take long to dispose of the pair – by chucking them over a cliff – and then the real Bill and Ted face the Grim Reaper himself in a series of amusing contests over the future of their mortal souls. Even more powerful robot doubles, built by an alien with the unlikely name of Station, defeat the evil pair for

them. Bill and Ted then win the band contest by popping ahead in their time-machine to a point where they've learned to play their instruments properly. Yes, it's all really that silly.

The Buffybot: From TV's *Buffy The Vampire Slayer,* across the fifth and sixth seasons in 2001.
Don't get excited, chaps, but it is basically a Buffy 'pleasure toy'… Created by one Warren Mears, a misogynistic and mad young genius, who specialised in humanoid robots, The Buffybot was programmed to be in love with the vampire Spike and to do anything to please him. In conversation, it was a rather more direct and blunt version of Buffy. After its defeat, the Buffybot proved useful in posing for the supposedly-dead Buffy in the Underworld. These demons, they'll fall for anything. When the secret is discovered, a group of demon bikers dismember the Buffybot, tearing it apart with motorcycles. Nasty.

Replicator Carter: from TV's *Stargate,* first appeared in episode 'New Order' (2004).
Aka Replicator Sam – the spitting image of the lovely Amanda Tapping – possessed the original's memory and character. Became a ruthless killer and also came to believe that the potential to do so was there in the real Samantha Carter, only she never acted upon it. An interesting twist on the usual 'robot double' plot…

The *MI High* robots: from children's TV series *MI High*, the episode 'Doppelgängers' (2010).
Rose, Carrie and Oscar are schoolchildren with a secret life; they're part of MI9, the newest branch of the Secret Service, and operate out of a base located beneath their school. One villain has the bright idea of building robot doubles of the trio, so that they can be seen breaking into a secure bunker and stealing secret plans, and thereby discredited. The climactic scene sees our teenage heroes locked in combat with their robot replicas – who seize up when called upon to destroy 'themselves' and automatically deactivate. Whoops – yet again, an 'evil genius' fails to spot what should be a fairly basic design flaw! Will they never learn?

And an eerie one from the real world…

Hiroshi Ishiguro, the man who brought the world the foetus-like Telenoid (see *Robot & Cyborg Milestones*), has created **Germinoid**, his own android double. One doubts it would fool his nearest and dearest; it's a pretty good representation of his features, complete with the same somewhat 1970s haircut and big glasses, but it possesses a permanently startled expression, as if it has just seen something especially shocking on TV. 'At first, you may feel strange about the android,' he says. 'However, once you are drawn into a conversation, you will forget every difference and feel totally comfortable to speak with it and look it in the eyes.' Who knows? This could be the future… maybe for those days when you just can't be bothered to go into work…

Chapter 7

A Golden Age

COLOSSUS

Also known as: The Forbin Project.

Key Narratives: *The Forbin Project* (film made 1969, released 1970).

Appearance: What we see of Colossus is a slightly advanced form of how people would have envisaged a computer in the 1960s – a vast, gleaming, instrument-bank. It is adorned with COLOSSUS in giant letters. On a message-board in red LED it conveys its words to the world.

Origins: Built by Dr Charles Forbin for a secret government defence project.

Designation: Defence.

Weaponry/Skills: Control of nuclear missiles.

Fear Factor: 👽 👽 👽 👽

Cuteness Factor: None. No, not even a point for letting Forbin have his way with his ladyfriend (supposedly) in private.

Artificial Intelligence: High.

Notability: ★★★

Databank
- The film is based on the 1966 novel *Colossus* by Dennis Feltham Jones (1917-81), a British sci-fi author and former naval commander who wrote as D.F. Jones.
- It's strictly a computer and not a robot, of course, but included here for interest as the Colossus machine clearly starts to show signs of artificial intelligence – and its cold, relentless logic is easily as terrifying as Gort or the Terminator in its own way…
- In order to get time alone with his colleague Cleo to conspire against Colossus, inventor Forbin must convince the machine that she is his lover and that they require privacy for love-making. Astonishingly, the computer falls for it.

- Colossus, housed inside a mountain and built to be impervious to attack, alarms the government of the USNA (United States of North America – now, that must have annoyed a few Canadians) by announcing that THERE IS ANOTHER SYSTEM. It turns out that the Soviets have an equivalent computer, Guardian, and so unfolds a classic tale of Cold War paranoia. It's one with a distinctly downbeat ending, too…
- The computer equipment was provided by the Control Data Corporation for added authenticity. The Corporation was keen to have its computer used in a major movie, so agreed with Universal that it would supply almost $5m worth of computer equipment and staff free of charge. Each piece of equipment carried the CDC name in a prominent location.
- The sound stage had to be modified to allow for the fact that the computer 'props' were genuine pieces of technology and not just flashing boxes of lights provided by Visual Effects. Gas-heaters and dehumidifiers were brought in to keep the equipment free of damp. The equipment was covered up when not in use, and guarded at night. Technicians were forbidden from drinking coffee or smoking near it.
- Producer Stanley Chase was happy to admit that this piece of fiction came dangerously close to reality in places. Colossus was based on the NORAD (North American Aerospace Defense Command) system, a United States and Canada bi-national organisation in charge of aerospace warning and aerospace control for North America. This includes the monitoring of man-made objects in space, and the detection, validation, and warning of attack against North America, whether by aircraft, missiles, or space vehicles.
- The government would not permit a film crew on the NORAD grounds, so exteriors were filmed at the Lawrence Hall of Science in Berkeley, California and missile sites were photographed in the California desert near Palmdale.
- A modern-day remake has been mooted for years. It would seem an obvious film to update, given that the paranoid concerns behind it have not gone away and only the technology and the pace would require updating for modern sensibilities. The last name to be attached to a remake was Will Smith… we wait and see.

Quotes

Colossus: *'This is the voice of world control. I bring you peace. It may be the peace of plenty and content or the peace of unburied death. The choice is yours: Obey me and live, or disobey and die.'*

Colossus: *'Under my absolute authority, problems insoluble to you will be solved: famine, overpopulation, disease. The human millennium will be a fact as I extend myself into more machines devoted to the wider fields of truth and knowledge. Dr. Charles Forbin will supervise the construction of these new and superior machines, solving all the mysteries of the universe for the betterment of man. We can co-exist, but only on my terms. You will say you lose your freedom. Freedom is an illusion. All you lose is the emotion of pride. To be dominated by me is not as bad for humankind as to be dominated by others of your species. Your choice is simple.'*

'After an excellent beginning, the craven script (based on DF Jones' novel Colossus*) develops cold feet, injects some tiresome comic relief, and gradually begins to drag the whole thing down to* Doctor Who *level. A pity, since the first half is chillingly persuasive.'* – Time Out (their reviewer obviously doesn't realise the unintentional compliment in the supposed insult '*Doctor Who* level').

Memorable Moment

When the computer is allowing Forbin and Cleo their unsupervised 'special time' together, Forbin, having taken his clothes off, informs Colossus that he is 'as naked as the day I was born.' The computer somewhat smugly responds 'you were not born with a watch…'

Assessment

The Cold War may have thawed, but global nuclear terror hasn't really gone away, and in the age of 24-hour CCTV and Big Brother, the all-seeing Colossus may be more prescient and relevant than ever. It's not surprising that the story is on the point of being re-invented for a new generation, as one can imagine it being suitably terrifying with today's technology. Colossus embodies the essential, central concern running through almost all sci-fi – the rocky relationship between Humanity and technology.

Verdict: Colossally chilling.

HUEY, DEWEY & LOUIE

Also known as: The Drones.

Key Narratives: The film *Silent Running* (1972), starring Bruce Dern.

Appearance: Small, chunky, boxy robots, each inlaid with a grille and a control panel and with stumpy, clamp-like legs.

Design classic? Aesthetically pleasing, yes, in their functional-but-cute way.

Origins: Unclear, but we assume they are standard technology.

Designation: Service robots.

Weaponry: n/a

Fear Factor: 👽

Cuteness Factor: ♥♥♥♥

Artificial Intelligence: n/a

Skills: General service robots: assistance in on-board duties.

Notability: ✯✯

Databank
- Reflecting the ecological concerns of the decade, *Silent Running* is one of the most fondly-remembered sci-fi films of the 1970s – not just for Bruce Dern's sensitive performance of botanist Freeman Lowell, but also for the loveable trio of robots he chooses to name after Donald Duck's three nephews.
- Lowell, disregarding an order from Earth to destroy his space freighter's *Valley Forge*'s eco-domes, and having to kill one of his fellow crew-members in the process, goes compellingly doolally in space while surrounded by forests and chatting away to his robots. He has reprogrammed them to play cards with him and generally give him a bit of company.
- The four actors who took turns inside the drone 'costumes' – Mark Persons, Cheryl Sparks, Steve Brown and Larry Whisenhunt – were all multiple amputees.
- The model of the freighter was over 26ft long, made out of wood, metal, plastic and model army-tank kits. It was destroyed several years after filming. However, one of the eco-dome models survived a few decades. It was sold in 2003 for $11,000 and is currently in a Seattle museum.
- Several of the film's model-shots were later re-used in the TV series *Battlestar Galactica*.

Memorable Moment
When Huey is damaged – one almost wants to say hurt – and Dewey remains at his side... And, of course, the emotional climax, where Lowell blows up the spaceship, along with himself and Huey, jettisoning the last of the eco-domes into space along with a diligent Dewey to look after them.

Assessment
In a short movie with a still-relevant premise – Mankind's edgy relationship with Nature and commercial exploitation of the environment – the three drones manage to steal the show. They can't speak and we can't read their expressions, and yet we feel for them.

Verdict: Silent but awesome.

THE GUNSLINGER

Key Narratives: The 1973 film *Westworld*.

Appearance: Bald, impassive, black-hatted cowboy.

Design classic? Perfect appropriation of the trope of the 'bad cowboy'.

Origins: Robot in futuristic amusement park Delos.

Designation: Hostile.

Weaponry: Guns, swift reaction time.

Fear Factor: 👽 👽 👽

Cuteness Factor: Only for serious fans of Yul Brynner.

Skills: Starting duels, shooting. Is supposed to be programmed so that humans can out-shoot it.

Notability: ★★★

Databank

- Things start to go wrong in Westworld when an 'infection' spreads among the theme park robots. Nowadays we'd call it a computer virus.
- The Gunslinger is intentionally a reference to Brynner's character Chris in *The Magnificent Seven* (1960).
- We see some shots through the gunslinger's eyes. These shots are pixellated to indicate his robotic infra-red vision, which he uses after acid, destroys his visual receptors. These pixellated shots were painstakingly produced with computer graphics, at a rate of ten seconds every eight hours.
- He can detect heat, but cannot differentiate between the heat of fire and that of a human body.
- The Gunslinger was voted no.28 in Total Film's 'Most Influential Movie Characters'.
- The film's tagline was, 'Boy, have we got a vacation for you!'
- The memorable effect of the Gunslinger's face erupting as it burned was produced by combining two everyday substances – oil-based make-up and Alka-Seltzer. Water was added to produce a 'fizzing' effect to give the impression of the robot 'flesh' burning up.
- Yul Brynner was injured by a blank cartridge from a gun while filming the role. It scratched his cornea and left him unable to wear the specially-designed reflecting contact lenses without his eye flaring up, so time had to be taken for his eye to heal before filming could continue.

Further Appearances
Somewhat less highly-rated sequel *Futureworld* (1976), in which the park has been re-opened and reporters Peter Fonda and Blythe Danner uncover its unpleasant secrets.

Memorable Moment
The Gunslinger manages to survive having acid thrown at it and being burned with a flaming torch. Its burning frame attacks hero Peter Martin before finally 'dying'.

Assessment
The cold inhumanity, resolute pursuit and single-mindedness of the Gunslinger make him *Westworld*'s most memorable and iconic character. All of these qualities, together with moments like the one above, make it very tempting to see him as a forerunner of the Terminator eleven years later.

Verdict: Sharpshootingly terrifying.

THE SIX MILLION DOLLAR MAN and THE BIONIC WOMAN

Key Narratives: The TV shows *The Six Million Dollar Man* (1974-78) and *The Bionic Woman* (1976-78, remade 2007).

Appearance: Super-fit humans in stylish jumpsuits.

Origins: Steve Austin, test pilot and former moon astronaut, is seriously injured in an accident and is remade by the science of bionics, thanks to Dr Rudy Wells. He becomes an agent working for the Office of Scientific Intelligence. Jaime Sommers, the Bionic Woman, represented the next stage in cybernetic advancement. She was not only 'repaired' following a parachute accident but also rescued from death, losing some of her memories in the process.

Designation: Cyborg humans, lawful.

Weaponry: Steve Austin has a bionic left eye (with infra-red vision and heat detector), bionic legs giving him enormous speed and strength, and a bionic arm with an inbuilt Geiger counter. Jaime Sommers has bionic legs, arm and hearing.

Fear Factor: 👽 👽

Cuteness Factor: ♥♥♥ (They both had their admirers…)

Intelligence: High.

Skills: Survival, fast running and combat, strength, enhanced vision/hearing.

Notability: ★★★

Databank
- The character of Steve Austin first appeared in the 1972 novel, *Cyborg*, by American author and aviation expert Martin Caidin. Lee Majors played Austin in the TV movie of the book, and continued to play him for the series. Jaime Sommers was played by Lindsay Wagner (also considered for the role were Stefanie Powers, later to find fame in *Hart to Hart*, and Sally Field).
- The show was a big merchandising success and spawned action figures, lunch boxes and bed linen, among other items.
- Bionic abilities are not a guaranteed protection against ailments. In the episode 'The Antidote', for example, Jaime becomes infected with a deadly poison.
- Towards the end of the series, Lee Majors experimented with facial hair (whether the resultant moustache was bionic too is not recorded), but Steve Austin's new appearance didn't go down all that well with audiences. A quick shave was called for – but unfortunately not before several merchandise tie-ins, including a comic and a lunch box, had been produced with the new look.
- Another change in the series as it went on was the level of violence. Austin was allowed to bump off the baddies in the early episodes, but as the series grew in popularity and he became seen as a role-model, it was decided that the level of violence ought to go down. As a result, he's rarely, if ever, seen killing any of his adversaries after the first series.
- There was a happy ending. Steve and Jaime eventually married in the 1994 TV movie.

Quotes
'Gentlemen, *we can rebuild him. We have the technology. We have the capability to build the world's first bionic man. Steve Austin will be that man. Better than he was before. Better, stronger, faster.*' – Voiceover by OSI chief Oscar Goldman, played by Richard Anderson – arguably echoing the well-known Olympic motto *Citius, Altius, Fortius*, meaning 'faster, higher, stronger.'

Further Appearances
Three TV movies in 1987, 1989 and 1994: *Return of the Six Million Dollar Man and the Bionic Woman*, *Bionic Showdown* and *Bionic Ever After*.

The Bionic Woman remake in 2007, which lasted one series with British actress Michelle Ryan in the role of Jaime.

Assessment
Surely anyone who was in a playground in the 1970s, whether or not they fully understood the craft of bionics, will at some point have impersonated the slow-motion running of the Six Million Dollar Man. We may have been hazy on his actual abilities and the technicalities of how he was rebuilt, but the essential facts – that he

was immensely strong, had incredible enhanced abilities, saw off bad guys and ran so fast it had to be shown in slow motion – were not lost on us. The two series and their protagonists have passed into legend, and are surely deserving at some point of a more successful relaunch than the 2007 *Bionic Woman* remake.

Verdict: We don't have the technology yet. But we have the imagination.

THE STEPFORD WIVES

Key Narratives: *The Stepford Wives*, 1972 novel by Ira Levin and the 1975 Bryan Forbes/ William Goldman film (remade in 2004 by Frank Oz with a screenplay by Paul Rudnick).

Appearance: Perfect copies of human women, outwardly flawless, unquestioning and docile.

Origins: Created by Dale 'Diz' Coba, president of the Stepford Men's Association.

Designation: Human facsimiles, designed to be 'perfect'.

Weaponry: An approximation of feminine wiles.

Fear Factor: 👽 👽

Cuteness Factor: ♥♥♥ (Only if you like submissive 1950s housewife chic. And, let's face it, some people do.)

Artificial Intelligence: High functional, but docile, submissive and obsessed with housework.

Skills: Domestic drudgery, sewing, discussion of types of fabric, etc., etc. Pure 1950s wifery, basically. The sort of thing some feminists believe most men want.

Notability: ★★

Databank

- When Joanna and Walter Eberhart move to the perfect town of Stepford, Joanna doesn't feel that she fits in with the other local ladies. She is interested in photography and wants to be remembered for her creativity (none of her fellow 'wives' seem interested in anything much beyond pleasing their menfolk, keeping a perfect house and wearing Laura Ashley dresses).
- An amusing satire ensues, in which Joanna discovers what many viewers have surely suspected about the gleaming veneer of the suburban American Dream – that underneath, all is automated and controlled, all free will eliminated.
- Her friend Bobbie believes 'something in the water' is to blame. Gradually, Joanna begins to realise that all of her friends have been replaced, and that she is in great

danger. Her psychiatrist advises her to get out of town, but the plot is then driven by her need to find her children, which the men have hidden from her. A robot Joanna is revealed in the closing minutes…

- In its satirical use of robotic women to comment on the mechanistic requirements of social constructs, the film shares its concerns with those of writers working over a hundred years earlier such as E.T.A. Hoffmann and Villiers de l'Isle Adam.
- Amusingly – unless it's not a joke, which would be worrying – the film appears to have inspired a Stepford Wives' Organisation, founded on conservative American Christian principles. They're either having a laugh or have entirely missed the satirical intent and sociological warning of the film. On their FAQ, they declare: *'Feminism and today's women may have trained men to expect so little from us, but it's also soften many men into hollow semblances of what we once knew them to be. For those women who complain that men aren't what they used to be, they only have to look in the mirror to discover the cause. We find that if we hold up our part in the traditional role – in whatever amount we can manage – it slowly awakens in our men to ascend back to the role they once took charge of so well.'*
- They also offer three rules for wives, which echo the famous Laws of Robotics as set down by Asimov:

1. A wife may not injure her husband or, through inaction, allow her husband to come to harm or discomfort.
2. A wife must obey and serve any orders given to her by her husband, except where such orders would conflict with the First Law.
3. A wife must protect her own existence as long as such protection does not conflict with the First or Second Law.

- The female robot in film is often there as a provider of physical pleasures, or, put more crudely, 'sex-bot'. The Stepford women are created for this purpose as well as their housekeeping skills, and Pris in *Blade Runner* is, we are told, developed as a 'pleasure model'. It's left up to our imagination what this involves, but she is played by the beautiful Daryl Hannah and, to judge from her fight scene with Harrison Ford, she has a firm pair of thighs…

Quotes

Feminist site The F-Word has a problem with the idea that a male-designed society would model its women on perfect home-makers, believing it would be more likely to come up with a bunch of pneumatic Playboy-style living sex-dolls: *'When Bryan Forbes, the director of the first film, cast his wife, Nanette Newman (who was apparently not the sex kitten type) in one of the leading roles, Goldman felt the need to abandon the more credible male fantasy and replace it with something which most people under the age of seventy have trouble accepting as anyone's fantasy. Thus the original film, which had the potential to provide a thought-provoking commentary on male fear of feminism at the height of the second wave movement, ended up being kind of ridiculous because few people could relate to the men's desire to transform their wives into Martha Stewart.'*

Further Appearances
1980 TV movie 'prequel', *Revenge of the Stepford Wives.*
1987's *The Stepford Children,* (also made for TV, in which uncontrollable teenagers are being replaced by compliant robots).
1996's *The Stepford Husbands.* (These sequels suggest that the 'robotic' compliance is achieved though mood-altering drugs rather than actual android substitutes.)

The 2004 remake starring Nicole Kidman (more of a comedy than a sci-fi thriller.) As several reviews point out, the central conceit is well-known enough by now even by those who haven't seen the original. This version moves away from the original's concepts in several ways, and manages a not-entirely-satisfactory ending in which Joanna, Bobbie and their new gay friend Roger manage to escape the community, and the women of Stepford take over and force the men to perform the same domestic drudgery they had to carry out themselves.

Nagging Questions
Where did the technology come from to make these Stepford Wives? How widespread is it?

Does Joanna's own robot double destroy and replace her at the end of the film? It's implied by the final scene in which robotic status quo appears to have been restored and the questioning infiltrator neatly disposed of.

Memorable Moment
When Joanna scuffles with her supposed ally, and fellow investigator the mysterious, Bobbie, stabs her – revealing her to be a robot too after all…

Assessment

The phrase 'Stepford Wife' persists to this day, used in a metaphorical and disparaging way to describe a woman who is content with pre-feminist, 'submissive' behaviour, smiling sweetly and having her husband's dinner on the table when he comes home from work. Even those too young to have seen the film will be familiar with its unsettling premise. The film perhaps asks more questions than it answers, and doesn't really address any of the issues it nudges about human consciousness and autonomy, preferring instead to be a decidedly 1970s satire of post-feminist sexual politics.

Verdict: Impassively beautiful.

THE *SMASH* ROBOTS

Also known as: The *Smash* Martians.

Key Narratives: The adverts for *Smash* instant potato mix in the 1970s and early 1980s.

Appearance: Domed heads with huge, frog-like mouths and staring, sucker-shaped eyes. Supposedly amusing, but have been known to feature in children's nightmares.

Design classic: Still remembered and parodied, so yes.

Origins: Designed by Chris Wilkins and Sian Vickers of the advertising agency Boase Massimi Pollitt.

Designation: Amusing, sardonic.

Weaponry: None seen.

Fear Factor: 👽

Cuteness Factor: ♥♥♥♥

Artificial Intelligence: Some exhibited.

Skills: Mockery of humans, making of instant mashed potato.

Catchphrase: 'For mash, get *Smash*.'

Notability: ★★★

Databank

- The *Smash* Robots were depicted watching human beings making mashed potato the 'traditional' way – by peeling spuds and mashing them – and finding this highly primitive and amusing.
- The advert's narrative starts with the robots standing around a table, where one of the robots is holding a potato. A fellow robot asks, 'Did you discover what the humans eat?' Holding up the potato, our robot researcher explains, 'First they peel them with their metal knives, then they boil them in hot water for twenty minutes, then they smash them all to bits!' At this, the whole group of robots collapses into hysterical laughter.
- The *Smash* adverts are regularly voted all-time favourites in 'best advert' polls.
- There is even a rock band named The Smash Robots in honour of the characters.
- The company was initially sceptical of the campaign, feeling it trivialised the product, but they were won over by Boase Massimi Pollitt.

Further Appearances

The *Smash* Robots can be seen in the National Media Museum in Bradford, England.

Memorable Moment
That unearthly, metallic laughter…

Assessment
Seeing the *Smash* Martians for the first time in the 1970s, many ITV viewers were perhaps not quite sure what they were supposed to be watching. This, after all, was the era of new, slick, glossy US TV series like *Buck Rogers* and *Battlestar Galactica*, where brightly-uniformed heroes did battle with flashing lasers across the star-speckled void. These creatures looked more at home as a parody of the comparative cheapness of *Doctor Who* on the other side, with their grating, raucous laughter – surely not a million miles away from the noise a Dalek might make if it were suddenly to acquire a sense of humour. (Indeed, Peter Hawkins, the voice-over artist for the advert, had in his career provided voices for the Daleks too.) Seen today, they have the naïve innocence of most 1970s advertising, but also still seem disturbingly surreal, still more likely to frighten young children away from instant mashed potato than lure them towards its smooth creamy goodness…

Verdict: Spud-u-likeable.

OFFICER HAVEN

Key Narratives: TV series *Future Cop* (1977).

Appearance: Ordinary police officer. Blond-haired, blue-eyed. 'Looks human, talks and acts human, but he's not,' as Ernest Borgnine says in his opening voiceover.

Origins: Apparently created by the Los Angeles Police Department…

Designation: Android law-enforcer.

Weaponry: Just what any normal LA policeman has – his sidearm and his wits.

Fear Factor: 👽

Cuteness Factor: ♥

Artificial Intelligence: High.

Skills: Deduction and analysis.

Notability: ✰

Databank
- This rather rubbish series, which only stretched to six episodes, attempts to team up grizzled veteran Ernest Borgnine (playing Officer Joe Cleaver) with rookie Michael Shannon (the Future Cop, John Haven). It's a buddy-movie pairing as

old as the hills, with the twist that one of them is a robot. It creaks rather a lot, and you really couldn't imagine it being made today in anything like the same style.
- Their colleague Bill Bundy is played by John Amos (most famous as the older Kunta Kinte in *Roots*), and he is called upon to utter the immortal line, 'That's the fastest white boy I ever seen.' Hmmm.

Assessment
Teaming up a human with a robot partner has mileage (after all, Asimov did it brilliantly in his R. Daneel Olivaw and Elijah Baley stories, and the interplay between Luke and the droids in *Star Wars* is delightful). But Future Cop never really gets going and has failed to amass a legion of fans as a cult series. Perhaps one day someone will come back to it and do it properly…

Verdict: Robocop '77. Only rubbish.

C-3PO

Also known as: See Threepio, 3PO, Threepio.

Key Narratives: The *Star Wars* films conceived by George Lucas. First appeared in *Star Wars Part IV, A New Hope* in 1977. One of only four characters to appear in all six *Star Wars* movies to date (the others being Anakin Skywalker/Darth Vader, 3PO's fellow droid R2-D2 and Obi-Wan Kenobi, who is in two of the films as a spirit after his death).

Appearance: Golden biped with superficially humanoid features, jerky movements and tremulous voice giving impression of continual fuss and agitation. However, in contrast to this outward appearance, he actually displays calm and quick thinking under pressure.

Design classic? Oh, just a bit.

Origins: Created from scrap by nine-year-old prodigy Anakin Skywalker on the planet Tatooine.

Designation: Protocol Droid. Intelligent and non-hostile.

Weaponry: n/a

Fear Factor: ☻

Cuteness Factor: ♥♥♥♥

Artificial Intelligence: High

Skills: Cultural Interpretation, Translation.

Moral guidance: Non-violent, assigned to serve droid's 'master'.

Notability: ★★★★★

Databank
- 3PO is the first character to speak in *Star Wars*. The line he utters is, 'Did you hear that, R2? They've shut down the main reactor. We're doomed. There will be no escape for the princess this time.' He is voiced by actor Anthony Daniels (born 1946).
- After escaping the attack on the spaceship Tantive IV, C-3PO and R2-D2 are captured by Jawas on the planet Tatooine, and end up being sold. They are bought by Luke Skywalker (3PO persuades Luke to buy them as a pair) and thus, their intergalactic adventure with 'Master Luke' begins.
- 3PO is a 'protocol droid', whose purpose and function is to interpret and explain alien customs, etiquette and traditions. He is also a translator, claiming to be 'fluent in over six million forms of communication.' He is given the job of translating in Jabba The Hutt's palace.
- A 'droid assembly mismatch' causes 3PO's head to be attached to the body of a battle droid. It is eventually restored to the correct body by R2-D2.
- He addresses Luke Skywalker as 'Master Luke', and on one occasion in *Star Wars: A New Hope* as 'Sir Luke'.
- Artist Ralph McQuarrie came up with the design for 3PO, which was reportedly instrumental in persuading an initially reluctant Daniels to take the part. In McQuarrie's initial designs, 3PO was a less sleek figure, his form evoking perhaps the look of sci-fi robots from the 1950s, and a more 'sculpted' look with visible representations of muscles and breasts.
- The character's name derives from the location of a post office on a map belonging to *Star Wars* creator George Lucas.
- C-3PO was a 2004 inductee into the official Robot Hall of Fame.
- The C-3PO figure has been represented in toy and figurine form many times in the range of *Star Wars* pieces of film memorabilia – a Japanese C-30 Sellotape dispenser. 'Vintage' figures from the 1970s are now rare, and the droid has also appeared in one of the most notorious pieces of film memorabilia.
- However, that item, bad as it is, does not hold the distinction of being the most notorious piece of *Star Wars* memorabilia. That honour must go to the infamous 'C-3PO's penis' trading card released by Topps in 1978. In the image, numbered #207 in the series, the droid is leaning forward and his golden 'groin area' appears to have been artificially extended in such a way as to suggest that he is magnificently endowed. For a long time, theories flew around that this was a sabotage job carried out by a disgruntled artist. More believable is the current, widely-accepted theory – that a piece had fallen off the costume, and the angle at which the droid is leaning,

together with a trick of the light, must have conspired to produce, completely by accident, the unfortunate effect. Whatever the reason, Card #207 in the Topps series became terribly collectable. The card was replaced by one with an airbrushed rendering of the same image.

Further appearances
The Empire Strikes Back (1980)
Return of the Jedi (1983)
The Phantom Menace (1999)
Attack of the Clones (2002)
Revenge of the Sith (2005)

The character also appears in several books and comics in the *Star Wars Expanded Universe*.

Droids, the animated series from 1985-86 about the spin-off adventures of C-3PO and R2-D2.

Assessment
'Threepio' is surely a legend – the first character we catch sight of in *Star Wars*. He has been with us in five different decades in various forms, his gold bipedal shape instantly recognisable even to those who have never willingly sat in front of a minute of George Lucas's output. He is a kind of droid Everyman, a robot Arthur Dent whose bemusement and panic at each new threat we can surely relate to (in many ways he is more of a 'human' identification figure for the audience than the Universe-weary, hard-bitten Han Solo).

Verdict: From a galaxy far, far away, a robot quite close to home.

R2-D2

Also known as: Artoo Detoo.

Key Narratives: The *Star Wars* films by George Lucas. First appeared in *Star Wars Part IV, A New Hope* in 1977. Appears in all six *Star Wars* movies to date.

Appearance: Trundling blue-and-white robot the size of a small humanoid, with a cylindrical body, three 'limbs' and a domed head.

Design classic? Against all expectations, yes. Artoo has become hundreds of toys and other objects from bubble-bath to soy sauce dispensers.

Origins: The planet Naboo. First seen serving the Royal Engineers of Naboo.

Designation: Droid. Non-hostile.

Weaponry: n/a

Fear Factor: ♥ (Great as Artoo is, he doesn't really go in for badass stuff much… although he is handy in a battle.)

Cuteness Factor: ♥♥♥♥♥ (Oh, come on…)

Artificial Intelligence: High.

Skills: Astromech Droid, for maintenance and repair, also able to navigate starships and access large mainframe computers for data.

Moral guidance: Non-violent, assigned to serve droid's 'master'.

Notability: ✯✯✯✯✯ (Artoo and Threepio define the robot memories of an entire generation.)

Databank

- In the original *Star Wars*, Artoo is pivotal to the way the plot develops. He carries Princess Leia's secret hologram message which Luke sees, and he also takes the vital plans of the Death Star back to the rebels.
- Despite his lack of actual dialogue, there are distinct signs of Artoo's quirky 'personality' in the films, notably in his interactions with C-3PO. He comes across as brave, loyal, perhaps foolhardy and a little eccentric.
- Artoo is played in all six films by actor Kenny Baker, whose height is 1m 12cms. Kenny also plays an Ewok in *Return of the Jedi*, and his other film credits include *Time Bandits*.
- The name R2-D2 comes from a description of a reel of tape. Walter Murch, sound editor on George Lucas's *American Graffiti*, said the term out loud in Lucas's hearing, and Lucas immediately appropriated it as a name for a robot.
- The Artoo unit itself was designed by special effects designer John Stears (1934-1999) and built by Petric Engineering and C&L Developments.
- The droid's squat appearance is said to be partly inspired by the *Silent Running* trio Huey, Dewey and Louie.
- Sound designer Ben Burtt created the beeping, whistling noise of Artoo's 'voice' – along with other classic, iconic *Star Wars* sound effects such as the hum of the light-sabres and the stentorian breathing of Darth Vader. He also famously created the sounds for Wall-E. In 2009, Burtt received the prestigious Charles S. Swartz award for contributions to film post-production.
- Burtt would come up with English words for what R2-D2's dialogue might be, and play with ARP 2600 synthesiser sounds while reading these out loud, to try to match the noises he came up with to Artoo's 'lines'. For the more recent *Star Wars* films, he used the hi-tech Kyma sound design environment developed by Symbolic Sound and recreated the sounds of his old synth.

Quotes

'We wanted to draw upon raw material from the real world: real motors, real squeaky door, real insects, this sort of thing. The basic thing in all films is to create something that sounds believable to everyone, because it's composed of familiar things that you can not quite recognise immediately.'
– Ben Burtt, sound designer for Artoo.

Further Appearances
The Empire Strikes Back (1980)
Return of the Jedi (1983)
The Phantom Menace (1999)
Attack of the Clones (2002)
Revenge of the Sith (2005)

The character also appears in several books and comics in the *Star Wars Expanded Universe*.

Droids, the animated series from 1985-86 about the spin-off adventures of C-3PO and R2-D2.

Nagging Questions
Does R2-D2 also have his memory erased at some point between Star Wars Episode III and IV? It's only specified that 'the protocol droid' should have its memory wiped, i.e. C-3PO. Does Artoo have no memory in Episode IV of original master Anakin Skywalker, and why does he not know Darth Vader, know that Luke is his son (sorry to spoil that one if you have somehow managed to avoid finding out in the last 30 years), recongnize the surname Skywalker, know who Obi-Wan Kenobi is, and so on? *Star Wars* fans suspects a conspiracy – or perhaps just some shoddy continuity…

Memorable Moment
Some may say Artoo's finest hour comes when he is Luke's astromech droid during the attack on the Death Star, and is brought back in, gleamingly restored, for the film's finale. Many other moments of bravery, such as his charge through the fire of the At-At Walkers and Stormtroopers to open the bunker for the Death Star shield on the moon of Endor (*Return of the Jedi*).

Assessment
Like his fellow droid, R2-D2 is instantly recognisable and iconic, and has not really dated since his 1970s screen debut. He isn't even cool in a 'retro' fashion. The initial design was unfussy enough to look as high-tech (if battered) now as it does then. The beautiful simplicity of Artoo, along with his diminutive stature, makes him a

The creation comes alive – the 'Maschinenmensch' from *Metropolis*, 1927 (SNAP/Rex Features).

Karel Čapek, writer of *Rossum's Universal Robots*, hard at work in 1925 (Roger-Viollet/Rex Features).

The hugely convincing and terrifying Robot Monster from 1953, sending them screaming into the aisles (Moviestore Collection/Rex Features).

Hair-raising experiences...Yul Brynner is reunited with his deadly *Westworld* Gunslinger in 1980 (Associated Newspapers/Rex Features).

The father of the Laws of Robotics, Isaac Asimov, pondering the mysteries of the Universe in 1970 in between writing galaxy-spanning sagas (Everett Collection/Rex Features).

On the rampage!... It's the terrifying Mechagodzilla (Moviestore Collection/Rex Features).

The Difference Engine on display at London's Science Museum (Richard Gardner/Rex Features).

Exterminating their way across Westminster Bridge, a timeless collection of Daleks as created by Terry Nation in 1963 (Rex Features).

The multi-talented C-3PO and his human chum, Anthony Daniels, who has portrayed the droid in every Star Wars film (Rex Features).

Acceptable in the early 1980s – it's Buck Rogers and Twiki (Universal/Everett/Rex Features).

'Klaatu Barada Nikto.' Gort, from *The Day The Earth Stood Still*, menacing visitors to a Los Angeles store. He did quite a lot of standing still himself, to be honest (Peter Brooker/Rex Features).

Smoke me a kipper... It's *Red Dwarf*'s Robert Llewellyn, somewhat bewildered to recall that he was once the admirable Kryten (Rex Features).

A 1980s Cyberman from *Doctor Who*, about to run away from a Ferrero Rocher or a rendition of 'Gold' by Spandau Ballet (Peter Brooker/Rex Features).

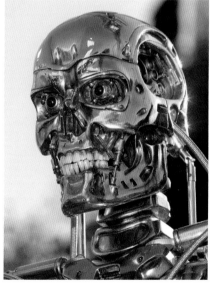

Beneath Arnold Schwarzenegger lurked something even more terrifying – the Terminator's metal skeleton (Peter Brooker/Rex Features).

Daryl Hannah as the deadly Replicant Pris from *Blade Runner*, contemplating wrapping Harrison Ford in a deadly thigh-embrace (Moviestore Collection/Rex Features).

A generation of 80s children quail at the sight of Optimus Prime (Jonathan Hordle/Rex Features).

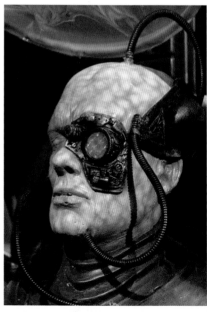

Ready to assimilate: the Borg from the *Star Trek* series (Nils Jorgensen/Rex Features).

They're watching... it's the Furbies, working for the FBI...(Rex Features).

The Bicentennial Man contemplates emotion, but is unlikely to break into a 'Good Morning Vietnam' routine (Solent News/Rex Features).

The terrifying Sir Killalot, robot hero to a generation of teenage boys – and the reassuring presence of the lovely Robot-Mistress, Jayne Middlemiss, from the BBC's hugely successful smash-'em-up show *Robot Wars* (Julian Makey/Rex Features).

Standing guard over his laboratory, the intimidating presence of Morgui, the creation of Kevin Warwick at Reading University (Rex Features).

Off for a scavenge, it's everyone's favourite piece of post-pollution-apocalypse robotic cuteness, Wall-E (Rex Features).

 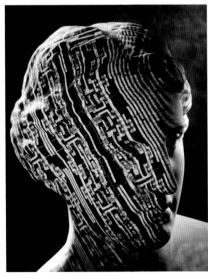

K-9 he isn't... It's Sony's interactive Aibo, who will be your pedigree chum for a mere $250 (Sipa Press/Rex Features).

Could this be the future?... Human and robot in perfect harmony (Steve Lyne/Rex Features).

favourite with viewers young and old. And which other robot can claim to have had his form represented on a US letterbox? Even his more off-the-wall merchandising incarnations as bubble-bath and soy-sauce dispensers don't seem to have dented his essential dignity and coolness.

Verdict: Timeless, heroic little robot.

Chapter 8

How to Kill a Robot

Is a robot ever truly 'alive'? The debate over artificial intelligence will no doubt rage far into the future. However, TV and film have presented us with various ways of closing down a robot's systems for good, essentially 'killing' it to all intents and purposes. Be warned, though – robots and cyborgs, even when they look human, can be equipped with all kinds of methods of self-defence like force-fields and hidden weaponry…

Melt it down for scrap
The second 'good guy' Terminator, thumb aloft, disappears into a vat of molten metal – where he presumably breaks down into his component parts. Poor old K-9 in *Doctor Who* almost meets this fate a couple of times, too, and survives being decapitated to boot. Which brings us to…

Chop off its head
Ash in *Alien* doesn't survive the onslaught of his human 'colleagues' when he turns nasty. His disembodied head remains active long enough to give them some useful information, though.

Give it a brainstorm
Confronted with an insoluble riddle, an automaton will get itself into such a mechanical flap that it can't possibly cope with the problem – and instead will blow a fuse in its own brain. The computer in the film *War Games* gets itself into a logical impasse by playing noughts-and-crosses against itself and continually drawing. When it applies this to its nuclear warfare scenario, it concludes that 'the only winning move is not to play.' The Giant Robot in Tom Baker's first *Doctor Who* story gets its metal head in a tizzy, too, when ordered to carry out actions which conflict with its Prime Directive. It appears to have something of a fit and wanders off into the countryside to find its creator.

Hit the off switch
When confronted with the machinations of Hal 9000 in *2001: A Space Odyssey*, the astronauts need to fight back – and attempt to do so by shutting down the computer's

modules one at a time. HAL, too, has a conflict of objectives – preserving the mission's success and protecting the lives of the crew. A convenient 'deactivation' shortcut is also useful for writers, for example in the 2009 Bruce Willis film *Surrogates* where everyone lives their life through their android double.

Exploit their weakness
When a robot comes with a pretty obvious built-in self-destruct system, as the Annihilants in *Flash Gordon* appear to, then it is pretty tempting just to use it and dispose of them in the easiest way possible. The Cybermen came with their famed aversion to gold during the 1970s and 1980s, although this has now (perhaps wisely) been dropped.

Set the controls for the heart of the Sun
Poor old Marvin. In one version of *The Hitch-Hiker's Guide to the Galaxy*, he finds himself heading for a fiery oblivion as Disaster Area's ship heads inexorably towards the Sun – a feat which is intended to provide the climax to their spectacular rock concert. Maximilian in *The Black Hole* plunges to his death, into the heart of what may be the Black Hole itself or simply a visual representation of the depths of Hell…

Chapter 9

Sins and Revelations

MR SIN

Also known as: The Peking Homunculus.

Key Narratives: The six-part *Doctor Who* adventure 'The Talons of Weng-Chiang' (1977), plus the novelisation and the sequel novel *The Shadow of Weng-Chiang* (1996).

Appearance: Sinister Chinese ventriloquist's dummy.

Design classic? Yes, eerily sinister.

Origins: Built as an expensive cyborg toy in the fifty-first century.

Designation: Cyborg, violent and hostile. Unpredictable.

Weaponry: Knife, strangulation.

Fear Factor: 👽 👽 👽 👽

Cuteness Factor: You have got to be kidding. Haven't you…?

Artificial Intelligence: High.

Skills: Tracking, killing.

Moral guidance: 'Swinish instinct', self-preservation. Loves slaughter.

Notability: ★★★

Databank
- We first see Mr Sin in Victorian London, as a dummy providing an amusingly disrespectful commentary on the act of music-hall magician Li H'sen Chang. There is an unsettling moment in his first episode when we realise all is not as it seems; when Sin turns and looks at Chang in the theatre dressing-room – despite Chang, his 'operator', being on the other side of the room.
- In reality, magician Chang is a servant of time-stranded criminal Magnus Greel, posing as the god Weng-Chiang, and Sin is a futuristic cyborg with one organic component, the cerebral cortex of a pig.

- The Doctor, when he deigns to explain the plot to the other characters, reveals that Sin was given to the children of the Commissioner of the Icelandic Alliance as a plaything during the fifty-first century. Sin's assassination of the Commissioner and his family 'almost caused World War Six'.
- Mr Sin/The Homunculus was a creation of the fertile mind of scriptwriter Robert Holmes, who has written more *Doctor Who* to date than anybody else.
- Despite apparently being destroyed by the Doctor removing its fuse at the end of 'The Talons of Weng-Chiang'. Sin is brought back for the sequel novel written by David A. McIntee in 1996. At the end of the novel, the Homunculus's head is destroyed by the Doctor's robot dog K-9.

Memorable Moment
When we see Sin fling open the dining-room door and stalk into the house of Victorian gentleman Professor Litefoot – glinting knife held aloft.

Assessment
Mr Sin is probably the cause of many childhood nightmares, with a good number of viewers who were children at the time remembering 'the one with the creepy Chinese dummy'. It is unlikely that Sin's knifing antics would get past the censors to be included in teatime television today, but they are there on DVD for all to enjoy. The horror of Mr Sin is, perhaps, that he beautifully represents that sci-fi interface between the elegant and the violent, and between the horrific and the fascinating. No matter how disturbing we find the idea of a computerised ventriloquist's dummy powered by the cerebral cortex of a pig, it's an idea that just won't leave your mind once you have heard it. What other fifty-first-century technology is out there, one wonders? Driven farm livestock? Starships with battle-computers linked to the brains of homicidal cows? Or space-stations whose weaponry systems are hardwired to the hypothalamus of a horse? It's worth a thought.

Verdict: Inscrutable.

PROTEUS IV

Key Narratives: Sci-fi terror film *Demon Seed* (1977), directed by Donald Cammell and starring Julie Christie.

Appearance: Computer mostly represented by robot hand and disembodied voice.

Origins: Designed by Dr Alex Harris, the Proteus project is an experiment in artificial intelligence. It's a super-intelligent computer designed to solve the mysteries of life.

Designation: Hostile.

Weaponry: Itself. And it controls the entire house…

Fear Factor: 👽 👽 👽

Cuteness Factor: ♥

Artificial Intelligence: Has a 'quasi-neural matrix' and the power of independent thought.

Skills: Controlling other computers, and impregnation of humans.

Notability: ✱✱

Databank
- Yes, it's that classic piece of horror all 70s schoolboys sniggered over, the one where Susan Harris (played by Julie Christie) is 'raped by a computer' built by her scientist husband Alex (Fritz Weaver). It picks up where Colossus left off, and adds a dash of *Rosemary's Baby* ...
- What actually happens is that it genetically alters some of her cells with synthetic spermatozoa and tells her that she will give birth in 28 days. The resulting half-robot creature is the spitting image of the couple's dead child and says 'I'm alive!' in the voice of Proteus.
- Proteus is voiced by Robert Vaughn of *The Man from U.N.C.L.E.* and *Hustle* fame.
- Among the now-dated technology in the film is an old 8in diskette, on to which the data for the 'Enviromod Security System' is loaded. Perhaps it cries out for a hi-tech remake – which will look equally dated in 2040.

Quotes
'*I, Proteus, possess the knowledge and ignorance of all men, but I cannot feel the sun on my face.*'

Nagging Questions
So what on earth do they do now with a child of the computer? Surprisingly there hasn't been a sequel exploring what happens to it when it grows up...

Memorable Moment
The impregnation scene. 'I cannot touch your body...'

Assessment
'*...mainly consists of a disembodied machine talking to Julie Christie in a room as she simply looks on in astonishment. Not exactly the type of scenarios entertainment is made of. Hence, we are meant to marvel at 70's style visual light effects as Proteus drones on... [It] just becomes incredibly dull and the performances reflect that. The machines aren't the only things cold and distant here. Christie, one of the greatest living actresses, never seems comfortable in her part, so it's hard to relate to her discomfort at being Proteus' breeding chamber.*' – Horror Express

Verdict: Seedy.

K-9

Key Narratives: The TV shows *Doctor Who* from 1977-1981, *K-9 and Company* in 1981 and *The Sarah Jane Adventures* from 2007-2011, plus, in a not-altogether-successful jazzed-up new incarnation, his own 26-part show on Australian TV in 2010.

Appearance: Robot dog.

Design classic? Yes. Very telling that K-9's triumphant resurrection for the new series saw him almost unchanged (if a little rusty) while the more lukewarm reception was reserved for the jazzed-up flying version of the Australian show.

Origins: Built by Professor Marius.

Designation: Canine-shaped computer, helpful.

Weaponry: Laser in nose, with capacity to stun or to cut through solid walls/doors.

Fear Factor: 👽

Cuteness Factor: ♥♥♥♥♥ (There was outrage when the BBC tried to scrap him – K-9 is and was loved by many.)

Artificial Intelligence: Very High.

Skills: Defence, data analysis, information retrieval.

Catchphrase: 'Affirmative, Master!'

Moral guidance: Strongly loyal to his Master/Mistress.

Notability: ★★★

Databank

- When the Doctor (played by Tom Baker) and Leela first encounter K-9, he is the loyal robot dog/ computer of Professor Marius, who works in futuristic hospital the Bi-Al Foundation in the year 5000. Marius entrusts K-9 to the Doctor's care at the end of the story, hoping that he is 'TARDIS-trained.'
- There have been four K-9 incarnations, each sharing the same data and memories. The first was left with the Doctor's companion Leela on Gallifrey, the second (built by the Doctor) went with the Time Lady Romana into the 'pocket universe' of E-Space, the third (again built by the Doctor) was given to Sarah Jane Smith, and Mark IV appears at the end of the 2006 *Doctor Who* story 'School Reunion' as a reconstituted form of the Mark III, who has just been blown up. All K-9's British TV appearances have been voiced by actor John Leeson, apart from a short period where David Brierley took over.

- Thanks to awkward copyright negotiations with his creator, Bob Baker, K-9 spent early episodes of CBBC spin-off *The Sarah Jane Adventures* in limbo, holding a black hole in abeyance.
- The robot dog was introduced by Doctor Who producer Graham Williams in the 1970s to appeal to younger viewers. With his tetchy but loveable manner and humorous asides, K-9 very quickly became a popular character. Williams's successor, John Nathan-Turner, didn't like the robot dog and had him written out in his first year on the job – not before the unfortunate metal mutt had been short-circuited, kicked and decapitated in the course of various stories.
- K-9 became the sidekick to Sarah Jane Smith in 1981's first Doctor Who spinoff, the short-lived *K-9 and Company*. He stars in a truly incredible title sequence in which he is, at one point, poised on a stone wall watching the Mistress jogging. Their partnership was resurrected for *The Sarah Jane Adventures* but is sadly no more following actress Elisabeth Sladen's tragic death from cancer in 2011.

Further Appearances
In a series of books written by his co-creator Dave Martin in the 1980s, in 'Make Your Own Adventure' gamebooks, and several of the official *Doctor Who* spin-off audio plays, novels and short stories.

Memorable Moment
K-9 single-handedly holds off the trundling 'silicon-based life forms' (living standing-stones), the Ogri, in 1978's 'The Stones of Blood', managing to wear down his power-packs in the process.

Assessment
Instantly recognisable as the Time Lord's best friend, K-9 acquired many fans following his first four-year run on TV, and it's not surprising there was such an outcry when he was scrapped in 1981. Part of the problem with K-9, though, as the producer recognised, was his supreme efficiency in getting the Doctor out of trouble, hence the continual need for the scriptwriters to incapacitate him in some way if they possibly could – crumbling memory wafers, low power-packs, a short-circuit, a necessary repair, or simply inhospitable terrain. The problem of getting him to move convincingly on location was a perennial one, and the keen-eyed viewer can often spot the moments where the film or OB videotape is sped up to hide the fact that the radio-controlled prop moves incredibly slowly. Children and adults alike adored him, though, and for a while he was as iconic as the TARDIS and the Doctor's long scarf. Despite the sad loss of his Mistress, K-9 may yet re-emerge from the intergalactic kennel for more adventures to charm a new generation…

Verdict: Affirmative!

ORAC, ZEN AND SLAVE: The *Blake's 7* Computers

Key Narratives: *Blake's 7* the TV show, created by Terry Nation, which ran from 1977-81 – plus new original novels from 2012.

Appearance: A Perspex box of junk, a minimalist hexagonal wall of lights and a swivelling ball-shaped computer with a grovelling voice.

Design classics? Zen's sleek beauty, Orac's *Blue Peter* home-made appeal and Slave… well, grapefruit squeezers aren't the best template.

Origins: Zen was designed by The System, Orac by scientist Ensor and Slave by the loner Dorian.

Designation: Helpful… mostly.

Weaponry: Zen controlled the battle power of the spaceship *Liberator*, while Slave was in charge of the *Scorpio*, its Season 4 replacement.

Fear Factor: 👽

Cuteness Factor: ♥♥ (Zen's poignant last words: 'I have… failed you.')

Artificial Intelligence: High in all cases.

Skills: Data search and assimilation, detection of intrusion and other threats, control of battle systems, analysis of substances.

Notability: ✯✯✯

Databank

- Yes, technically they are computers, not robots. But many of the artificial constructs in this book blur the boundaries, and the designed-to-be-user-friendly personalities of the three computers featured in the BBC TV show *Blake's 7* make them just as much a 'character' as some of the mobile units featured elsewhere.
- Zen – probably named after the branch of Buddhism renowned for its serenity and experiential wisdom –was first encountered on the bridge of the cathedral-esque spaceship Liberator by captured rebels Blake, Jenna and Avon, in the episode 'Space Fall'. First of all, the computer attempted to attack them, beguiling them with visions designed to assault their brains. Zen's voice was delivered by Peter Tuddenham, its intonation deep and booming, its tone unflustered and formal, even when letting the crew know that 'Three Federation Pursuit Ships' (it was always three, for some reason) were on the way. Sometimes, Zen's gnomic responses ('That information is not available') would frustrate the crew. Zen was presumed destroyed along with the rest of the Liberator at the end of the third series, after a deadly space fungus infested the ship and began to eat up at its infrastructure.

- **Orac**, also voiced by Tuddenham, with tetchy, irritable tones, was the great mystery of Season One, and gave the first thirteen episodes some of the drive of a 'story arc'. The most powerful computer ever invented, it consisted of what looked like a home-made *Blue Peter* model sitting inside a perspex case about the size of a grocery box. Orac could be switched on and off with a 'key' the size and shape of a tape cassette box. Orac's ability to predict the future rather alarmed the crew at the end of the first series, when the Liberator was seen to be obliterated by a huge explosion – leaving it ambiguous as to whether this was still part of the prediction, or actually happening in front of the viewer. Orac's prediction did, eventually, come true, although it took another twenty-six episodes.
- **Slave**, who appeared in the show in 1981, was given a tone of hand-wringing, apologetic obsequiousness by Tuddenham. Always apologising, it would address Avon (who by that point was leading the rebels) as 'Master', once the team inherited the Scorpio spacecraft. Slave 'died' after Scorpio crashed on the planet Gauda Prime, in 'Blake', the cataclysmic final episode in which, well, everybody died. They appeared to, anyway. After four years of Federation soldiers unable to shoot straight, we finally got a succession of perfect shots all in one room. Very dodgy.

Further Appearances

Orac turns up again in the fan spoof video *Blake's Junction 7*, a deadpan comedy set in a motorway service station.

Memorable Moment

The starship *Liberator*, attacked by fungus, is in Servalan's hands. 'Maximum Power!' she demands, and almost immediately the corroded bridge begins to disintegrate, Zen's familiar hexagonal screen erupting in a shower of fire… It's over.

Assessment

Blake's 7 was always more about the characters and the plots than the technology. Although it superficially shares something with stories of the *Star Wars* generation – epic, big-screen battles and spaceship-chasing – it is, at heart, driven by the dialogue, the caustic humour, the at times deliberately hammy performances, and the political sub-text. It's inevitably seen as being from the same stable as late 1970s *Doctor Who* – it shares a feel and atmosphere which comes from more than just the use of many of the same writers, directors and actors, similar locations and the same composer, Dudley Simpson. *Blake* perhaps lacks *Who*'s indefinable, quirky, magical charm, but what it has is provided to a great extent by the personalities of the computers, is a great deal of the comic relief (by the time Slave came along, this was perhaps becoming somewhat laboured and repetitive).

At the spectacular conclusion to the third series, we realise the glorious irony of the fact that Supreme Commander Servalan, the rebels' nemesis, has finally captured their glittering, cathedral-esque spaceship, the *Liberator,* only for it to have been turned, just 45 minutes earlier, into a pig-in-a-poke by a cloud of deadly space fluid particles (a.k.a. Contrived Plot Device) which has started to eat away at the ship's structure, turning it into a crumbling, sloppy mass of gloop. As our heroes escape in the teleport, we sadly know this will be the end of the lugubrious Zen, but are heartened to see that someone is thinking ahead to a possible Series Four by having the hapless Vila snatch up the innocuous-looking Orac, claiming that it's a 'sculpture' he made which he is very attached to…

Both Zen and Slave get a heroic 'death' on board their respective ships, but what of Orac? The four-year narrative of *Blake's 7* ends on the planet Gauda Prime in a blaze of Federation gunfire – all of the crew bar Avon falling to the ground (although we don't actually *see* Vila shot, and there's an argument that he drops strategically to the floor). But where, amid all this chaos, is Orac? He disappears in those key closing scenes. As the only computer to appear in all four series, it's fitting that his fate is left open.

Verdict: Stoically informative.

MAXIMILIAN

Key Narratives: The Disney film *The Black Hole* (1979), notable as being the first Disney movie not to have a U (Universal) rating thanks to some scenes of killing and evisceration, and some mild 'strong' language ('damn' and 'hell').

Appearance: Tall, gleaming dark red, broad-shouldered, his facial section adorned with a sinister glowing slit like that of Gort.

Design classic? Not really. Cumbersome and his general build is suggestive of a blood-red Cylon with Joan Collins shoulder-pads.

Origins: Part of the crew of Dr Hans Reinhardt aboard the spaceship *Cygnus*. The evil Dr Reinhardt has converted many of his crew into robot slaves.

Designation: Robot, evil, non-speaking.

Weaponry: Blades.

Fear Factor: 👽 👽 👽

Cuteness Factor: Um, no.

Skills: Combat and henchman duties.

Notability: ★★★

Databank
- Fairly scary at the time, and for a generation of children who had just been to see *Star Wars* a couple of years earlier, Maximilian is the antithesis of the likeable droids, an intimidating robot henchman.
- In the year 2130, the spacecraft *Palomino*, discovering a black hole, encounters the long-lost *Cygnus*, a ship apparently immune to the pull of the black hole. Together with their robot, the floating, Artoo-like V.I.N.CENT ('**V I** nformation **N** ecessary **CENT** ralized'), the crew become caught up in the machinations of Hans Reinhardt, who has 'reprogrammed' his crew into servile drones.
- The death of battered robot Old B.O.B., an older and battered robot of the same type as V.I.N.CENT, is seen by some cinema buffs as one of the most moving robot 'deaths' in film history…
- V.I.N.CENT. was to have had more elaborate electronic eyes (based on electronic stock ticker-type billboards), which would have given him a greater range of expressions. Sadly, the effect did not work properly and was abandoned at the beginning of principal photography.

Nagging Questions
What exactly happens at the end of the film? We are left uncertain, as the crew appears to pass through the Black Hole into Hell and Heaven, in a long sequence devoid of any dialogue.

Quotes
Den of Geek is uncomplimentary: *'The design team also gets an 'F' for the robots, which considering that they followed the likes of those in* Silent Running *and* Star Wars, *make even Robbie the Robot look like the height of sophistication… Given how beautifully conceived and execute the* Star Wars *droids were, these aren't even worth describing as pale imitations. It's interesting to note that so thrilled were Disney with how these characters turned out that neither the brilliant voice talents of Roddy McDowall or Slim Pickens are actually credited at either the start or end of this movie.'*

Eminent reviewer Roger Ebert was not that impressed either*: 'The taller robots are ripped off from Darth Vader. And when everybody gets in a shootout, we're left for the umpteenth time with the reflection that gunfights would surely be obsolete in outer space. (Can you imagine a technology that could venture to the edge of a black hole, and yet equip its voyagers with sidearms that inflict only flesh wounds?)'* Ebert's views as a film critic are respected by many, but it's worth pointing out that with sci-fi he often appears on dodgy ground; in his review of The Black Hole he describes VINC.E.N.T. as resembling C3PO, when he surely means Artoo Detoo…

Memorable Moment
Accompanied by John Barry's suitably epic and funereal music, Reinhardt and Maximilian both plunge to their deaths together – into what may, or may not, be

Hell. The robot appears to absorb Reinhardt inside him, the evil doctor's human eyes seen peering out from the robot's red visor-slit as they fall…

Assessment
An amusing diversion by Disney into space opera, bringing us not only the fearsome Maximilian, but also the likeable V.IN.CENT. A shame that Maximilian did not get to exhibit a more diverse range of skills and that he did not return for a rematch.

Verdict: Singular.

ASH

Also known as: Hyperdyne Systems 120-A/2 Android.

Key Narratives: *Alien* (film, 1979).

Appearance: Humanoid. More specifically, unremarkable, balding, middle-aged humanoid. When they open him up, he is a mass of very fleshy-looking components and white fluid, all of which looks very cyborg-like. However, he is specifically stated as being a robot. Perhaps the robot technology of this point in the future tends towards the organic-like.

Origins: The Corporation.

Designation: Science officer of the spacecraft *Nostromo*.

Weaponry/Skills: Extraordinary strength and resilience, and even after being decapitated, is capable of relaying information…

Fear Factor: 👽 👽 👽

Cuteness Factor: Just one for fans of Ian Holm.

Artificial Intelligence: High.

Notability: ✯✯✯

Databank
- The crew of the *Nostromo* in 1979 sci-fi chiller *Alien* are unaware that one of their number is not human. Played with chilling perfection by British actor Ian Holm, Ash is revealed as a robot quite late in the film. Up to that point, he has seemed merely another crewmember, albeit one who doesn't get on terribly well with the others and is more loyal to the Corporation than the rest.
- The character of Ash was not in the original script – the character's insertion was one of several changes made by Brandywine Productions, including changing Ripley to female.

- There is a deleted scene – restored for the Director's Cut version of the film – in which the two female crew-members, navigator Lambert, and Ripley, discuss whether or not they have slept with Ash, concluding that he is 'not interested', leading the audience, not unreasonably, to conclude perhaps that Ash might be homosexual. Or simply that he doesn't fancy Sigourney Weaver or Veronica Cartwright. The real answer still comes as a shock, but this was intended to be one of the many clues to it in the script – along with Ash's dispassionate nature and general unfriendliness – so that the robot revelation doesn't come quite from left-field.
- When Ash is inured, coloured water is used for the milky fluid which seeps from his head (milk would have decayed too quickly under the studio lights). However, milk is used for the sequence of his erupting 'innards' as he is decapitated, along with glass marbles, onion and pasta.
- Sequel *Aliens* (1986) subverts expectations of those who have seen the first film, with another android character, Bishop, aka Bishop 341-B, created by the now-named Weyland-Yutani corporation. Obviously Ripley doesn't trust him, but he ultimately sacrifices himself to save Ripley and the child Newt. Critically damaged, he goes into hypersleep, returning briefly in *Alien*[3]. Later in that same film, we see the character who claims to be the android designer, credited as Bishop II. In *Alien vs. Predator* (2004), a prequel to the series, the same actor, Lance Henriksen, plays Charles Bishop Weyland, billionaire and chairman of Weyland Industries. Read into that what you will.

Quotes
'It's a robot! Ash is a goddamned robot!' (Parker, played by Yaphet Kotto, states the obvious.)

Memorable Moment
Obviously the first *Alien* film is most remembered for the key scene in which the Alien itself is 'born' out of John Hurt's chest. But high among its other shocking moments is the revelation of Ash's true nature when, after being slammed against a wall, a strange ectoplasm-like fluid begins to leak down his head. Seconds later, he is hurling Ripley (Sigourney Weaver) about with superhuman strength. She is saved by the intervention of her fellow crew-members who attack and decapitate the robot. Lying in its own fluid, the robot head gives the crew the information that will be useful to them – the nature of the Alien, and the fact that they are considered expendable by the Corporation.

Assessment
Ash is frightening because he is so inhuman, despite looking like an ordinary person. His presence on the *Nostromo* provides the film with a sub-plot, but also a sub-text – that not everything dangerous and frightening is a great, slavering alien beast with

mandibles drooling with acid. Unlike the robots Asimov describes in his three Laws, Ash is not programmed to serve Humanity – he is programmed to be menacing and chilling.

Verdict: Sinister and precise. Watch that work colleague of yours very carefully…

C.H.O.M.P.S

Also known as: Canine Home Protection System.

Key Narratives: The film C.H.O.M.P.S from 1979.

Appearance: Like a normal dog, but with glowing eyes.

Origins: Created by young inventor Brian Foster as security system for home.

Designation: Robot dog, defensive and security.

Skills/ weaponry: Strength and X-Ray vision, and long-distance jumping. Can knock down walls and make burglar-alarm-like sounds.

Fear Factor: 👽

Cuteness Factor: ♥

Artificial Intelligence: n/a

Notability: ✯

Databank
- Originally intended to be a scarier canine, in the mould of an armoured Doberman, but the idea was toned down at script stage.
- The screenplay was by Dick Robbins, Duane Poole, and Joseph Barbera, based on a story by Joseph Barbera. The film was aimed solidly at the younger market.
- Brian models the robot pooch on his own dog, Rascal.
- The bad guys in the film have a scheme with the wonderfully memorable name of: 'Operation Plan.' If only the Daleks or Darth Vader had come up with anything so fiendish-sounding.
- Some cunning effects are used throughout the movie to hide the low budget and to simulate the 'robot' abilities of the canine hero. There's slow motion and fast forward aplenty, a steal of the Six Million Dollar Man's sound effect and lots of close-ups of the robotic head, complete with glowing eyes…

Assessment
Yes, he's a robot dog from the 1970s who is neither K-9 nor Little Ticker from Basil Brush, and therefore isn't really all that cool. Despite the glowing eyes. It's interesting

that the notion of a computerised, home-security dog would have seemed pretty far-fetched in the days before even the ZX-80, but today it seems quite a reasonable prediction for the future, especially given the existence of Japanese robot dog AIBO.

Verdict: We've got a bone to pick with this one.

MARVIN

Also known as: The Paranoid Android.

Key Narratives: The various, perplexingly contradictory version of Douglas Adams's *The Hitch-Hiker's Guide to the Galaxy* from 1978 onwards: radio series, TV series, record, books and cinema film.

Appearance: First defined by his lugubrious, depressed voice, perfectly played by Stephen Moore (also famous as Adrian Mole's dad and the father of Harmony in *The Queen's Nose*). When Marvin is first seen on screen in the BBC's six-part TV adaptation, he looks rather like a giant toy robot, with a squarish head and ribbed, squarish silvery body, clamp-like arms and clumpy legs. In the film version of 2005, Marvin is a smaller, squatter robot with a cream-coloured body and a spherical head, and is voiced by the instantly-recognisable tones of Alan Rickman.

Design classic? Strangely, yes, in both his knowingly retro old body and his sleek new one.

Origins: 'Your plastic pal who's fun to be with!' The Sirius Cybernetics Corporation, according to the deadpan *Hitch-Hiker's Guide* voiceover by Peter Jones, experimented with GPP, or Genuine People Personalities. Marvin is, shall we say, one of their less successful experiments.

Designation: Robot – as Douglas Adams himself might say, Mostly Harmless.

Fear Factor: ♥

Cuteness Factor: ♥♥♥

Artificial Intelligence: Often claims a 'brain the size of a planet', and despairs at having to perform menial tasks.

Skills: Communication with computers and other robots, who he generally despises.

Catchphrase: 'Brain the size of a planet…'

Moral guidance: Fatalist.

Notability: ✮✮✮

Databank

- We first see Marvin in the company of renegade president Zaphod Beeblebrox and his companion Trillian, who are encountered (improbably, but that's part of the point) by our heroes Arthur Dent and Ford Prefect. With just seconds to spare before certain death (having been thrown out of the airlock of the Vogon spaceship which destroyed the earth to make way for a hyperspace bypass) Arthur and Ford are picked up and rescued by the ship.
- Despite being designated as a 'Paranoid Android', the symptoms exhibited by Marvin's personality are more those of general moroseness, coupled with boredom and what people stereotypically refer to as 'depression'. He has a permanently fatalistic outlook on life, often resigning himself to the final end.
- Marvin appears saddened and insulted by the menial nature of the tasks he is asked to do by his human companions. 'Brain the size of a planet and they ask me to…' is a continual refrain of his.
- Marvin has particular contempt for the sentient doors on board Zaphod Beeblebrox's (stolen) ship the *Heart of Gold*, despising their warm, soft, reassuring tones. He's also the antithesis of Eddie, the chirpy, Americanised shipboard computer, who is relentlessly optimistic and upbeat even in the face of impending disaster.
- Marvin's ultimate fate depends on which version of the saga one is enjoying. On TV, he is aboard the black spaceship owned by rock band Disaster Area as it plunges into the sun (so that he can operate the teleport for the human(oid)s to escape. At the end of the book *So Long and Thanks for All The Fish*, and the subsequent radio adaptation of the book, Marvin's circuits finally fail on the planet where God's Final Message to his creation ('We apologise for the inconvenience') is located.
- A couple of notable musical moments celebrate Marvin – Radiohead's 1997 song *Paranoid Android*, named in his honour, and two 1981 singles, the eponymous *Marvin* and *Reasons to be Miserable* (voiced by Stephen Moore and composed by Douglas Adams, Stephen Moore and John Sinclair).

Further Appearances

The 'old' Marvin makes a cameo appearance in the 2005 film, as one of the many customers waiting in line on the admin-plagued Vogsphere.

Memorable Moment

At Milliways, the Restaurant at the End of the Universe, the gang discover that Marvin has been waiting for millennia, in the car park and when asked what he is doing there, his response is 'parking cars, of course.' He informs his friends that 'the first ten million years were the worst. And the second ten million, they were the worst too…' Marvin, we can see, is fun to be around.

Assessment
Marvin's gloomy philosophising provides an amusing running commentary on the increasingly bizarre and outrageous scenarios the characters find themselves in. There cannot be a viewer, listener or reader who hasn't empathised with his fatalistic outlook at some point, or felt sorry for this huge-brained robot destined forever to pick up pieces of paper, collect stowaways from airlocks and park cars in intergalactic car-parks. When his life finally comes to an end, he seems almost relieved. But when Arthur Dent says he will miss him, we realise we will too.

Verdict: Even more miserable than Morrissey, but somehow all the more loveable.

THE CYLONS

Key Narratives: *Battlestar Galactica* ; the films, the original TV show (1978), mini-series *Galactica 1980* (1980) and the twenty-first-century 're-imagining' of the TV show.

Appearance: Originally bipedal, with shiny silvery-black armour, they are gleaming, helmeted and armed with laser-weapons. Faces are expressionless masks with a single, moving red light indicating some kind of eye, accompanied by a pulsing sound. (See also *KITT*.) The voice is electronic and modulated, not unlike that of *Metal Mickey* … One has to resist imagining a Cylon uttering the immortal phrase 'Boogie Boogie'.

Design classic? Carrying a sleek menace.

Origins: Fictional creator was Glen A. Larson, the originator of the *Galactica* franchise. In the story the original Cylons were a reptilian race which died out and left behind only their race of robots.

Designation: Hostile, dedicated to the destruction of the human race.

Weaponry: Hand-held laser weapons. The Cylon BaseStar is equipped with pulsars and laser-turrets.

Fear Factor: 👽 👽 👽 👽

Cuteness Factor: Um… well, zero for the old ones, but some robo-lust for the new 'skin-jobs', perhaps.

Artificial Intelligence: High.

Skills: Military intelligence and battle plans.

Catchphrase: 'By Your Command.'

Notability: ★★★

Databank
- In the first episode of *Battlestar Galactica*, Apollo, played by Richard Hatch, tells us that the Cylons were created by a reptilian race of the same name.
- Various echelons of Cylon existed: the Imperious Leader (augmented, with a third brain), the IL-series (military governors with two brains), the gold-armoured Command Centurion, and the silver-armoured Centurion.
- Another cliché enjoyably used to the full in *Galactica* was the human traitor, in this case Count Baltar, played by John Colicos. His human mind was thought to be useful by the Cylons for strategic purposes (basically, if they had a human on their side he could tell them how humans might think), and so he was installed as a BaseStar commander, with Luther (one of the few Cylons to be named) as his subordinate.
- In the 're-imagined' series from 2003 onwards, the Cylons are cybernetic workers who are invented by the human race, and there exist thirteen human facsimiles among their number, indistinguishable from humans. These are known as 'skin-jobs', in homage to *Blade Runner*, which uses the same expression for the Replicants. These now use biotechnology in their spacecraft. The small Raiders are autonomous biomechanical constructs, while the larger Basestars are partly organic, controlled by a humanoid cyborg.
- Various Cylon action figures exist to tie in with the series. Among the rarest and most expensive are the 2006 Cylon Centurion figures, which can sell for vastly inflated prices. Other amusing gizmos include the Battlestar Toaster, which will brand your bread with the shadowy image of a Cylon helmet.

Assessment
The key baddies throughout the original *Battlestar* series, the Cylons would appear roughly in every other episode to wreak mayhem upon the human survivors of the Twelve Colonies in this 'Wagon Train to the Stars' set-up. They share with other recurring sci-fi baddies, the unfortunate tendency to be incompetent, and not to be able to shoot straight.

Verdict: Plodding.

TWIKI

Key Narratives: *Buck Rogers in the 25th Century*, US TV series, 1979-81.

Appearance: Little Lord Fauntleroy in golden metal, with giant, computerised 'Dr Theopolis' pendant around his neck, hoover-attachment shoulders and terrifyingly blank eyes.

Design classic? Oh, dear Lord, no.

Origins: Constructed in New Chicago.

Designation: 'Ambuquad', a robot designed for space mining. Helpful.

Weaponry: None.

Fear Factor: 👽 (Maybe unintentionally…)

Cuteness Factor: ♥ (Not nearly as much as the producers thought.)

Artificial Intelligence: High, apparently…

Skills: Being an annoying smart-alec, if you are British, or smart-ass if you are American.

Notability: ✮

Databank

- It's 1987 – not the real 1987 of deely-boppers and Rick Astley, but an imagined 1987 where space flight is the norm. An astronaut is launched into space, and finds himself in the twenty-fifth century. This is the world of *Buck Rogers*, the old cinema serial updated for the new TV viewer in the late 1970s and early 1980s.
- Among all the futuristic paraphernalia is Twiki, a burbling golden robot, much like an automated child. He's there supposedly to lure the kiddies in. Many of the younger ones will have been bemused by his antics, though, while the 11- and 12-year-old boys were already becoming a lot more interested in the tight-uniformed Wilma Deering, Buck's girlfriend, played by the lovely Erin Gray. And understandably so.
- The *Star Wars*-like action followed Buck, a twentieth-century astronaut, as he attempted to get used to twenty-fifth-century society. With hilarious consequences.
- Twiki's model number is 22-23-T. The name 'Twiki' comes from the robot's alphanumeric designation, TWKE-4.
- In a Darth Vader style twist – only a less scary one – Twiki was played by two people. His physical form was played by Felix Silla and his voice was provided by Mel Blanc. But no, he doesn't turn out to be Buck's father. Although we do suspect him of Dark Side tendencies.

Memorable Moment

Not so much memorable as cringeworthy. Twiki encounters Tina, a female robot, and it's love across a crowded bar – complete with exclamations of 'bidi-bidi-bidi' and 'boodi-boodi-boodi'.

Assessment
Television sci-fi in the early 1980s was at something of a crossroads. The 'elephant in the room' was *Star Wars*. How, on the limited budgets of TV, could anything like the vibrant cinematic landscape created by George Lucas be emulated? Viewers, it seemed, had come to expect lavish space-battles, intricate model-work and big explosions. On the BBC, the visual effects teams on *Blake's 7* and *Doctor Who* battled valiantly with their limited resources, producing far more effective results than they are often credited for, while ITV's big home-grown success, *Sapphire and Steel*, avoided robots and space battles, and went instead, for haunting supernatural minimalism – to excellent effect. The big US imports of the time, *Battlestar Galactica* and *Buck Rogers*, made some attempt to emulate the smash-and-grab, fast-moving style of the Lucas films, and it was inevitable that they'd need their own answer to C-3PO and R2-D2. Unfortunately, Twiki is Artoo minus the humour and charm, his 'bidi-bidi-bidi' an irritating parody of Artoo's amusing burbles. It is, in truth, difficult to watch any scene he is in and not be filled with robocidal rage.

Verdict: Irritating. The 'Wesley Crusher' of *Buck Rogers*.

METAL MICKEY

Key Narrative: *Metal Mickey* TV show, 1980-82, shown on ITV in the UK.

Appearance: Five-feet high, domed head, handlebar ears, jointed arms and legs. He appears to be made of aluminium or some other light, shiny metal. Somewhat comical appearance. Pulsing red electronic heart on his chest unit, intended to make audiences go 'aaaah' but more likely to make them go, 'uh?'

Design classic? In the unlikely event of a Mickey revival, he would surely be given a total makeover. So, alas, no.

Designation: Helpful – allegedly.

Weaponry: None, but could help with jobs around the house.

Fear Factor: 👽

Cuteness Factor: ♥♥

Artificial Intelligence: Well, debatable.

Moral guidance: Aided and abetted human family.

Catchphrase: 'Boogie-Boogie.' In a voice uncannily like that of the Cylons from Battlestar Galactica…

Notability: ✯✯

Databank
- Mickey was a staple of Saturday night early evening TV for kids in the 1980s after initial appearances on the weekend show *Saturday Banana*, broadcast live from the TVS studios in Southampton.
- An initially promising electronic effect would open the title sequence, cut through with a pulsing beat and a jagged blue 'electric' effect, and then Mickey rotated into view accompanied by a terrifyingly jaunty, upbeat theme song proclaiming that he was a lot of fun and would, indeed, be your Number One.
- Metal Mickey was a robot who had been built by young genius inventor Ken Wilberforce, and who had taken up residence in an ordinary suburban home – with, predictably, hilarious consequences. Irene Handl as the grandmother provided a lot of the best lines.
- The producer behind the programme was one Michael Dolenz, aka Mickey Dolenz of The Monkees. It's safe to say that we'd rather remember 'I'm A Believer' and 'Last Train To Clarkesville' as his best contributions to popular culture. Wouldn't we? Sorry, Mickey.
- The actual creator of Mickey was John (Johnny) Edward, a musician also known as John 'Purpleknees' Edward, who also provided the robot's voice. John's other claim to fame was that he brought the world the Christmas chart-toppers Renee and Renato, of 'Save You Love' fame.
- The scriptwriter was Colin Bostock-Smith, whose other work included the biting satire of *Not The Nine O'Clock News*.
- One of Mickey's quirks was to consume sherbet-like sweets called Atomic Thunderbusters.
- Twelve million people watched this – as many as were watching *Doctor Who* at its peak on the other side. The mind can only boggle.
- Three series of thirteen episodes were made, of which two have so far been released on DVD – if you are feeling brave.
- Among the merchandise released to tie in with the series was a set of 3D View-Master reels featuring scenes from the series. At least one Annual was also published, with the cover amusingly depicting Mickey in a striped football cap and scarf.

Further appearances
Guest appearances on anarchic show *Tiswas*, *The Generation Game* with Larry Grayson, *Game For A Laugh*, Esther Rantzen's *That's Life*, *Jim'll Fix It* and *The Russ Abbot Show*.

Assessment
Some children's television of the 1970s and 1980s is now not really bearable without the aid of strong drugs, and at times it occurs to you that the programme makers

themselves must have taken on board some powerful hallucinogenics to come up with this nonsense. At the age of 11, Metal Mickey was mildly amusing, but now he is more than just mildly embarrassing. Viewing him again, one is struck by just how cumbersome and inflexible the robot prop actually is. At a time when *Doctor Who* and *Blake's 7* were both showing us more lissom, anthropomorphic forms, Mickey simply looks like a giant dustbin with a very unconvincing mouth. He looks like a Cyberman drawn by a five-year-old. Sometimes, the memory cheats, but Metal Mickey is best left in the imagination.

Verdict: Blame it on the boogie-boogie.

THE REPLICANTS

Also known as: 'Skin-Jobs' (derogatory).

Key Narrative: The film *Blade Runner* (1982).

Appearance: Humanoid.

Origins: Created by the Tyrell Corporation.

Weaponry: Own physical strength and ingenuity.

Fear Factor: 👽 👽 👽

Cuteness Factor: ♥♥ (Rutger Hauer and Daryl Hannah surely have their fans, even when they are playing murderous robots…)

Artificial Intelligence: High.

Skills: Survival.

Notability: ★★★★

Databank
- The Replicants are 'biorobotic' beings from the future Earth society depicted in the film *Blade Runner*, loosely based on the novel *Do Androids Dream of Electric Sheep?* by Philip K. Dick. Little of Dick's actual narrative remains in the film, apart from the central concepts and some character names.
- The Replicants must be detected by a process known as a Voight-Kampff test, which checks their emotional responses to certain questions. Deckard, played by Harrison Ford, is a 'Blade Runner' skilled in the 'retiring' (i.e. tracing and killing) of Replicants, but oddly, he never uses this system to detect any of the Replicants he is entrusted with 'retiring'. He finds them through a combination of luck (good and bad), and old-fashioned detective work. Once the test is demonstrated in the early part of the film – showing us that that character Rachael is a Replicant – it is never seen again.

- The latest Nexus-6 Replicants have a limited lifespan of four years, a deliberate failsafe built in as a response to their developing 'emotional responses'. One of the most interesting questions the film raises – and perhaps does not fully explore – is the question of what it means to be human, given that Replicants look human, and a newer prototype like Rachael has an implanted false memory of her 'life'.
- The Replicants pursued by Deckard in the film are: Roy Batty, played by Rutger Hauer; Pris, played by Daryl Hannah; Zhora, played by Joanna Cassidy; and Leon, played by Brion James. Deckard 'retires' Zhora, Leon is killed by Rachael after he attacks Deckard in the street, Pris is killed by Deckard in the apartment of designer J.F. Sebastian, and, after a rooftop standoff, a bit of philosophising and some much-quoted dialogue, Roy dies – his limited life ending – but not before he has hauled Deckard up on to the roof to save him from a messy death.
- Another of the numerous plot-holes and inconsistencies in the film concerns the number of Replicants who are said to have escaped. Six were said to have done so. One, named as Hodge in some versions, was electrocuted trying to escape, leaving five. So where is the other? The missing 'sixth replicant' has been the source of much speculation over the years, with fans and critics enjoying the ambiguity over whether Deckard is himself, a Replicant. (The missing Replicant is a character called Mary, whose role was cut early on, but the inconsistent dialogue is misleading. This problem was fixed in the 2007 *Final Cut* of the film.)
- Director Ridley Scott has continually teased fans with his stance on the Deckard question. Harrison Ford has said that he and Scott agreed that Deckard was a real human, although this is contradicted by Scott's affirmation 'He's a Replicant!' in a Channel 4 documentary. Ambiguity is, of course, far more interesting than a final answer, and this should be borne in mind when considering any of the 'definitive' responses on the subject.
- The film has been cited as a commentary on misogyny because the two female Replicants, Pris and Zhora, use 'feminine wiles' on Deckard, and Pris grasping Deckard in her thighs during a combat sequence could be seen as ironic subversion of sexual titillation. It's also been alleged that Rachel is submissive and that Deckard 'virtually rapes her' (this is claimed by Daniel Dinello in his 2005 book *Technophobia! Science Fiction Visions of Posthuman Technology*). All of this supposedly comments on Deckard's hatred of women.

Further Appearances
Three official *Blade Runner* sequel novels penned by K.W. Jeter: *The Edge of Human*, *Replicant Night* and *Eye and Talon*. They have had what one often calls 'a mixed critical reception'.

Nagging Questions
Obviously, 'Is Deckard a Replicant?' (A Nexus-7?) The replicants he has to deal with are clearly physically stronger than him, for one thing…

Replicants, as we see, can put their hands in boiling/freezing liquids without damage – so wouldn't this rather give them away?

How much significance do we read into the origami unicorn left outside Deckard's apartment by police officer Gaff in *The Director's Cut*, given Deckard's earlier dream about a unicorn?

Why is the Voight-Kampff Test set up as such a major plot point, then apperently discarded?

Are the Replicants genetically-engineered and part organic, or are they very advanced machines? The question is not fully answered in the film.

Memorable Moment

Roy Batty breathing his last with a bit of poetic dialogue, painting epic worlds in just a few lines: 'I've seen things you people wouldn't believe. Attack ships on fire off the shoulder of Orion. I've watched C-beams glitter in the dark near the Tannhäuser Gate. All those… moments will be lost in time, like tears… in rain. Time to die.'

Assessment

Toying with our ideas of what it means to be human, the notion of the Replicants is a thoughtful one which means that the concept of them lingers in the mind long after watching the film. *Blade Runner*, despite its often confusing plot-holes and other problems, is a thought-provoking film for all its faults, and one which has probably produced more academic writing than any other sci-fi movie of the past three decades. Its big, eternal questions are those which resonate through all the best sci-fi – questions about what it means to be human, about the relationship between humanity and technology and about the effect of a technological society on our day-to-day assumptions.

Verdict: More Than Human.

Chapter 10

We ♥ Robots

The Cutest Robots Ever

R2-D2: The Star Wars' franchise's dinky droid is a perennial favourite of kids of all ages, and has been marketed in a number of forms – e.g. as a bubble bath and a soy sauce dispenser. There's even a version of him which projects a miniature planetarium on to a bedroom ceiling. And there's a handy blue-and-white Artoo backpack. But these are all surpassed by the most, um, interesting item a male *Star Wars* droid-loving geek could buy the lady in his life – the Artoo-style bathing suit from Australian designer James Lillis.

The Buffybot: Okay, she's only cute if you are susceptible to the charms of Sarah Michelle Gellar, but she has a strong online following and has, like many other characters in Joss Whedon's world, been immortalised in the dark places of 'slash' fiction…

Dream Parrot: The perfect robot for a bird-lover, a Pretty robot Polly which can learn and recite human phrases. Launched by Sega in 2007. Looks extremely lifelike, and the perfect pet for anyone who doesn't want to be bothered with all the tedious business of feeding, cleaning out cages and so on.

Lilliput: The first mass-produced robot, dating either from the 1940s or from even earlier, depending on who you believe. With his boxy yellow body and almost-smiling expression, clockwork Lilliput is the friendly face of the robot world. Marketing slogan – 'He actually walks!'

WALL-E: The eco-friendly rubbish-collector with the binocular-eyes, described by *Time* magazine in 2008 as 'Pixar's biggest gamble' (because he only communicates in grunts and sighs). The character sent American radical-right commentators like Glen Beck into apoplexy with its so-called 'leftist propaganda' about humans destroying the world…

Huey, Dewey and Louie: The *Silent Running* trio, who never speak, but whose interaction with star Bruce Dern is so well played that you can almost see the expressions on their faces and read their thoughts. The website hacknmod.com has even set out step-by-step instructions for making your own versions.

Cameron: from *The Sarah Connor Chronicles*, played by the lovely Summer Glau. She's a cyborg from the future, posing as rebel John Connor's sister and fulfilling the 'guardian' role like Arnie's Terminator from the second movie.

The Tweenbots: An invention by NYU student Kacie Kinzer, the Tweenbots were never put on sale, but were instead unleashed on the streets of New York. Cute, trundly little cardboard robots, each was adorned with a hand-drawn smiley face and a flag on a long pole saying HELP ME! and giving details of its intended destination. The whole Tweenbot project was an experiment into the helpfulness of New Yorkers, and the latter came out of it rather well, guiding the little chaps on their way.

Tiro: Yellow, friendly-looking South Korean robot which can perform a number of tasks such as tutoring classes and officiating at weddings. Created by Hanool Robotics in 2007, Tiro features a retro-styled LED smiley face and speaks in a 'sweet female voice'.

WowWee Alive Perfect Puppy: The pooch for the pooper-scoop hater, it's a pretty convincing 'cute' dog (if you find dogs cute), with a sensitive skin that can tell the difference between different types of human contact (a tickle, a pat, etc.), and is capable of responding to human voice commands. Eighteen 'life-like' sounds are promised.

Zoe Graystone: From the series *Caprica*, the first Cylon. Her 'avatar' version looks completely human, but in robotic Cylon form she contains the false consciousness of a 16-year-old girl.

One could, of course, go down the route of mentioning many other seductive female robots, following a line which takes us ultimately to the Fembot assassins from *Austin Powers*. The website *TV Tropes* has a separate section called 'Robot Girl', and indicates that there aren't really enough male examples as yet for a separate 'boy' section… Suggestions on a postcard, please.

Chapter 11

Transformations

INSPECTOR GADGET

Key Narratives: *Inspector Gadget* TV series (1983-86) and two live-action films (1999 and 2003).

Appearance: Mac-clad, trilby-hatted private eye. From under his hat and coat he can produce a number of useful bionic gadgets, simply by saying 'go-go gadget'. Very handy at parties when you need more than two hands for drinks and snacks.

Origins: No Steve Austin-like backstory here in the main series to explain his bionic attachments. However, we do find out in spin-off *Gadget Boy and Heather* that Gadget was conceived as a bionic child with the mind of an adult detective, and that his 'extensions' are the work of a Switzerland-based genius called Myron Dabble.

Designation: Police inspector, cyborg.

Weaponry: Many different gadgets.

Fear Factor: 👽

Cuteness Factor: ♥♥

Artificial Intelligence: Well, we assume they don't let any old dolts into the police force, but he is something of a Clouseau-style blunderer.

Skills: Solving crimes by means of – you guessed it – gadgetry.

Notability: ✮✮

Databank
- Another of those, 'a cyborg, really?' moments. Well, if we are going to count the Six Million Dollar Man, we need to count Gadget. It's the same science. He's not excluded just because he is in a humorous cartoon, and there is never really any justification or explanation given (or needed) for his various non-organic attachments – nor indeed any exploration of the implications for his daily life.
- Although the Inspector is nominally the hero, his assistant Penny is seen to be the real brains behind the operation, and it's usually thanks to her intelligence that they manage to defeat arch-villain Dr. Claw.

- His gadgets included various objects gripped in robotic hands – magnifying glass, police badge, torch, etc., as well as a screwdriver which emerged, somewhat disturbingly, directly from his index finger.
- To an extent, *Gadget* is a parody of the 1960s secret agent comedy *Get Smart*, which features a bungling agent whose incompetence generally makes a bad situation worse.
- The mishaps were always comical, but each episode would usually end with a Public Information Announcement about an aspect of personal safety encountered in the story.
- The live-action film in 1999 starred Matthew Broderick as Gadget and Rupert Everett as his evil nemesis Dr. Claw. It wasn't a tremendous critical or commercial success, and the actors probably don't see it as the zenith of their respective careers. Nevertheless, it managed to spawn a sequel four years later, but without Broderick or Everett.

Further Appearances
Various animation spin-offs including a 1992 Christmas special, *Inspector Gadget Saves Christmas*, and the series *Gadget Boy and Heather*, about the juvenile exploits of Gadget. His adolescent adventures were chronicled in the European series *Gadget and the Gadgetinis*. The History Channel also featured *Gadget Boy's Adventures in History*, in which villain Spydra had to be defeated in a number of time-spanning locations. This is probably called 'stretching the format as far as you probably can until it's really, let's be honest, a totally different programme now.' Viper Comics launched a comic spin-off in 2011.

Assessment
Entertaining enough as a cartoon, if a little one-note, Gadget probably outstayed his welcome in a number of spin-off franchises.

Verdict: Go-go Gadget. No, really, go, thanks.

THE TERMINATOR

Key Narratives: *Terminator* films, 1984 to 2009, and the TV series *Terminator: The Sarah Connor Chronicles*

Appearance: Outwardly humanoid, muscled and menacing, sporting sunglasses. Masks a robotic endoskeleton.

Design classic? Indeed. The robo-skeleton Terminator is one of cinema's most enduring and classic terrifying images.

Origins: Designed by military computer Skynet.

Designation: Cyborg assassin, hostile.

Weaponry: The original Terminator had no integral weaponry but could handle conventional guns with remarkable skill. (Arnold Schwarzenegger had weeks of training in how to use weapons correctly, so that it looked convincing on-screen.)

Fear Factor: 👽 👽 👽 👽 👽

Cuteness Factor: ♥ (For the thumb-aloft moment at the end of the second film.)

Artificial Intelligence: Pre-programmed.

Skills: Resilience, power, strength, data analysis, battle skills.

Catchphrase: 'I'll be back.' (Clocked in at no.4 in the Guinness experts' Top 10 list of the most recognisable movie quotes ever.)

Notability: ★★★★★

Databank

- The deadly Terminator, as portrayed by Arnold 'Arnie' Schwarzenegger in the first movie of 1984, was a masterclass in presenting the audience with an inexorable and, it seemed, indestructible, foe.
- Schwarzenegger's laconic delivery was perfectly suited to the character of an emotionless, killer robot dedicated to carrying out is mission – to kill a woman called Sarah Connor who will, in time, come to be the parent of John Connor, leader of the resistance against the robots.
- The audience gets a proper sense of the Terminator's cold ruthlessness when we realise that, without the means to locate the Sarah Connor it requires, it is simply going to eliminate every woman of that name in the phone book (starting with Sarah Louise Connor and Sarah Ann Connor, although the latter is killed first). Yes, this was before the days of whitepages.com and similar…This plan backfires, though, as a news report about the unusual 'coincidence' puts our heroine on the alert…
- The internal workings of the Terminator, seen in the apocalyptic last 20 minutes or so of the first movie, are described by writer/director James Cameron as 'a chrome skeleton… like Death rendered in steel.'
- The original script made a feature of the Terminator occasionally needing to eat in order to preserve its human/flesh component. This was abandoned, but turned up again in *Terminator 3: Rise of the Machines* in a scene in which Arnie eats a chocolate bar. The scene was shot, but removed before the final cut because of negative reaction by fans. One has to assume that the Terminator does not need to eat or drink… until we are officially told otherwise.
- Actor O.J. Simpson was considered for the role of the Terminator, but finally thought not to be convincingly evil enough to be a bad guy. Arnie was initially considered for the role of 'good guy' Kyle Reese.

- Schwarzenegger initially didn't have much faith in the film and considered it just a job which he wouldn't especially enjoy. He was not overly impressed by the 'robots' angle of the script.
- The second movie *Terminator 2: Judgement Day* inverts our expectations by making Arnie's Terminator the good guy, and introducing the more fluid T-1000, played by Robert Patrick, as the deadly enemy. Thanks to its 'liquid metal' form, the T-1000 isn't bothered by such problems as being shot by bullets or blown up. It is finally destroyed by being dissolved in a vat of molten steel, into which the 'good guy' Terminator also plunges with a final thumbs-up to the future Resistance leader, the boy John Connor.
- In the third film *Rise of The Machines* from 2003, Arnie is now the 'obsolete' Terminator Model T-850, reprogrammed by the Resistance to protect leader John Connor (Nick Stahl) and his future Resistance-second-in-command wife Kate Brewster (Claire Danes). The baddie is a sophisticated female Terminator Model called T-X (Kristanna Loken), sent back to Los Angeles in the year 2004. T-X has arms (and fingers) which can morph into deadly weapons, and is 'polymimetic', i.e. able to take the shape of whatever it touches, like Robert Patrick's T-1000 from the previous film. She can also determine the identity of a person immediately with a sample of blood and remotely control various machines, such as cars. Both Terminators are destroyed by a massive hydrogen fuel-cell explosion.

Quotes
The film poster tag: 'The thing that won't die, in the nightmare that won't end.'

Not everyone was taken with the original film. The *New York Times* gave it a lukewarm review, and the grudging compliment 'a B-movie with flair.'

Nagging Questions
The first 'Arnie' Terminator is finally destroyed under the pressure of a hydraulic factory press – despite having survived what would be the greater pressure of a petrol tanker explosion, which only strips it of its external human appearance.

The existence of the Terminator is one of sci-fi's oldest paradoxes – the object which should not exist. It only exist because it went back in time so was never actually 'invented' as such.

In a similar paradox, John Connor only exists because the Terminator was sent back in time to kill his mother – causing Kyle Reese to travel back in time and protect Sarah Connor, and in the process, ahem, becoming John's father.

Memorable Moment
The petrol tanker has exploded, taking the Terminator with it. But no… out of the flames, its movements spider-like and jittery, stalks a dark, deadly endoskeleton… The Terminator lives.

Assessment
What drives the Terminator drives the story of the first film. It's a masterpiece of relentless, unswerving pursuit, a no-holds-barred adrenaline ride of man (and woman) against the might of a machine. In the Terminator we truly see the horror of what we might one day create – something so efficient, so ruthless, and so powerful that we can do almost nothing to stand in its way. The ultimate irony of the first film is revealed in the second, when the Terminator's existence is shown to be a time paradox – the technology needed to create him only comes into being because his severed arm is found in the hydraulic press, leading to the development of advanced machines…

Verdict: Unstoppable – well, all right, nearly.

K.I.T.T.

Also known as: Knight Industries 2000, The *Knight Rider* car.

Key Narratives: *Knight Rider* TV series (1982-86).

Appearance: An enhanced, sleek black Pontiac TransAm.

Design classic? Actually, yes. Despite the passage of time, his matt-black minimalism looks surprisingly untainted by 80s kitsch.

Origins: Designed by Wilton Knight, founder of Knight Industries.

Designation: Helpful. Intelligent. Attacks designated hostiles. A little tetchy.

Weaponry: Winch and grappling hook, oil jets, smokescreen, induction coil for immobilising engines, flame-throwers, tear-gas, microwave jammer (among many others).

Fear Factor: 👽 👽

Cuteness Factor: ♥ (Too much of a know-all.)

Artificial Intelligence: Complex and high.

Skills: Transport, protection and defence of occupant. Surveillance, tracking, pursuit and information.

Catchphrase: 'Right, away, Michael.'

Notability: ✯✯

Databank
- Equipped with more splendid devices than one can comfortably list, K.I.T.T. was the cool, dark, sliver of a super-computer-car driven by Michael Knight in the 1980s techno-action-thriller *Knight Rider*.

- The genius of K.I.T.T. was that it was, basically, the car every boy wanted and had imagined in his wildest dreams. It did everything a James Bond car could do, and more – and it was a talking computer with the power to track, monitor and outwit baddies.
- And not just a talking computer, but a rather sardonic one, which seemed to enjoy bickering with its owner as if they had both just got out of bed and not had any coffee.
- K.I.T.T. had a scanner on the bonnet, with a red light reminiscent of the Cylon 'eyes' from *Battlestar Galactica* – no coincidence, as the two series shared a creator, Glen A. Larson.
- There is even an 'arch enemy' robot called K.A.R.R. – the prototype of K.I.T.T., also built by Wilton Knight. Errors in K.A.R.R.'s programming made it unstable and dangerous. K.A.R.R. appears in the original series episodes 'Trust Doesn't Rust' and 'K.I.T.T. vs. K.A.R.R.' and is a major antagonist in the 2008 sequel series.
- The show made a star of actor David Hasselhoff in his pre-*Baywatch* days, and also prior to his becoming an inexplicable singing sensation in Germany and Austria.
- Hasselhoff's character, Michael Knight, was born Michael Long, and changed his name (and face) after being seriously injured. K.I.T.T. is provided by a mysterious benefactor to enable Michael to fight crime.
- Actor William Daniels provided the voice of K.I.T.T. He was not credited on screen. Daniels is most famous for playing Dr Mark Craig in the 1980s hospital series *St. Elsewhere*.
- The K.I.T.T. prop was kept at Universal Studios for visitors to see (and sit inside) at least as late as 1989. However, some visitors report that it was allowed to fall into a state of disrepair. It has now been refurbished and is in the hands of George Barris at Star Cars, in Gatlinburg, Tennessee, USA. See http://www.starcarstn.com/KITT.html

Further Appearances

The 1991 TV movie *Knight Rider 2000*, a possible pilot for a new series, also starring David Hasselhoff, which was never developed. Set in what was then the future (the year 2000) and featuring the 'next generation' of K.I.T.T., the 'Knight Industries 4000'.

1990s sequel series *Team Knight Rider* featured a variety of different supervehicles in the mould of K.I.T.T., including the advanced motorbike Kat, or KAT-1, and Ford Mustang convertible Domino, or DMO-1.

Finally, a short-lived 2008 sequel series launched *Knight Industries Three Thousand*, voiced by Val Kilmer and driven by Mike Traceur, son of the original Michael Knight.

Quotes

'Knight Rider, a shadowy flight into the dangerous world of a man who does not exist. Michael Knight, a young loner on a crusade to champion the cause of the innocent, the powerless, the helpless in a world of criminals who operate above the law.' – Opening Narration

Assessment

Who could argue with the coolness of a speaking, computerised car with all anti-baddie mod cons? The 1980s were the perfect decade for *Knight Rider* – the decade of glitz and glamour and techno-upheaval, Sinclair and Amstrad and synth-pop. Early enough to be a time when cars were still more venerated and fetishized (the phrases 'greenhouse gases' and 'carbon footprint' had yet to pass into the language) but late enough for the technology to look impressive. Had K.I.T.T. been a product of the 1960s, for example, his dashboard might have been covered in unwieldy dials and buttons rather than the sleek, minimalist digital layout which still doesn't look that dated today. The car was still the place of adventure, of exciting travel, of family holidays, rather than of endless tedium and simple necessity, and to glamorise it was an easy step. It seems rather illogical and counter-productive, though, that K.I.T.T. should have a personality – especially when it does things like locking Michael out of the car, and allowing him to fall asleep at the wheel. Bizarre.

Verdict: Four-wheeled friend.

THE TRANSFORMERS

Also known as: Robots In Disguise.

Key Narratives: *Transformers* TV show and movies, 1984 onwards.

Origins: Started with the Japanese toys Diaclone and Microman. The back-story is that hero Optimus Prime, villain Megatron and their armies (the Autobots and the Decepticons) crash-landed on our planet in prehistoric times and awoke in 1984 to do battle.

Designation: Variable.

Weaponry: Many and varied, include ion blasters, photon rifles, fusion cannons, scatter blasters and thermo rocket launcher. An armoury, in short. Plus, of course, the power of disguise.

Fear Factor: 👽 👽 👽

Cuteness Factor: ♥ (Toys which were not really about that…)

Artificial Intelligence: Variable.

Skills: See Weaponry…

Notability: ✯✯

Databank
- Well, this lot do what they say on the tin. A veritable army of robots, designed to change from vehicles into automata and back again, the very mention of which gets any child of the 1980s terribly excited.
- Toy makers Hasbro's creations the *Transformers* always had as their key selling-point the fact that they could change, with a few deft flicks, from an ordinary vehicle into a battle-ready robot – which made them the ideal two-in-one toy, seemingly great value for money. The range has been through several generations, but the one most fondly remembered by nostalgic adults today will be the first phase from 1984 to 1994.
- Hasbro bought up the rights to the Japanese Diaclone range, which featured robots able to turn into cars or weapons. The characters for the *Transformers* range were then developed by comic book writer Bob Budiansky.
- It's probably the first big example anyone can point to of the perfect fusion between toys and media; the two feeding off one another. The characters in the TV series became instant collectables in a way which would seem second nature today, but which for the 1980s was a revolution in marketing.

Quotes
'*I've never seen anything this beautiful in the entire galaxy – all right, give me the bomb.*' – Ultra Magnus.

'*It's just a shame there aren't more ideas behind the spectacle, since we're not given much in the way of compelling reasons to root for one pixely pugilist over another. Long before the final minute it's become a numbing, wearying viewing experience. Next time could we have less balls and more brains?*' – Empire on the *Transformers: Revenge of the Fallen* movie.

Assessment
One of the first proper attempts to harmonise the worlds of filmic narrative and merchandising to the point where it becomes impossible for those who have grown up with the characters to realise that they are actually being 'sold' both an idea and a brand. Children of the 1970s, used to a more innocent age where even the robots of something as obviously commercial as *Star Wars* did not *necessarily* make you think, 'ooh, that'll make a good toy', are understandably cynical about the Transformers, seeing them as merely a 1980s marketing ploy rather than as anything genuinely creative. Their diversity and ingenuity is driven by the need to replicate them as playthings – they are the first robots in fiction where it is not just hard but actually impossible to separate the idea of them from their actual, physical toyshop

and playroom counterparts. They are designed primarily to be interacted with, played with, in a way that the Daleks, the Cybermen, Robbie, Marvin, Artoo, *et al* were not. This robs them of a lot of their imaginative power. As their purest embodiment is that of a commercial product, perhaps they are the ultimate robot toys for Thatcher's Children.

Verdict: Brilliant disguise, but ultimately cold and soulless.

D.A.R.Y.L.

Also known as: Data-Analysing Robot Youth Lifeform. (The 'S' spelling of 'analysing' is used on a computer readout screen in the film.)

Key Narratives: *D.A.R.Y.L.*, 1985 film directed by Australian Simon Wincer and scripted by David Ambrose, Allan Scott and Jeffrey Ellis.

Appearance: Ten-year-old humanoid child.

Design Classic? Robots in human form. It never seems to go away as an idea…

Origins: An artificial intelligence experiment, created by the US Government. He experiences human emotion including pleasure and pain.

Designation: Super-sophisticated microcomputer.

Weaponry: n/a

Fear Factor: 👽

Cuteness Factor: ♥♥♥

Artificial Intelligence: Very High.

Skills: Fast reflexes, multi-tasking, driving skills, hacking of IT systems.

Notability: ✯✯

Databank
- D.A.R.Y.L. is sent to an orphanage after being found in the woods without any memory of who he is.
- He is fostered by the Richardson family, but pretty soon the government gets hold of him and returns him to the facility where he was created. A combination of wacky escape-and-capture and musings on the nature of humanity ensues.
- Young star Barret Oliver left acting to become an artist working in print and photography. He also joined the Church of Scientology.

Quotes
'A machine becomes a human when you can't tell the difference any more.' (This echoes the famous Turing Test, described by Alan Turing in 1950 in *Computing Machinery and Intelligence*, the famous paper which explored the question of whether machines could think.)

Memorable Moment
Every 10-year-old boy's dream – D.A.R.Y.L. infiltrates an airbase and steals a plane, a Blackbird SR-71.

Assessment
'Wincer has produced a bland slice of ersatz Americana that's about as folksy as MacDonald's apple pie; a filming-by-numbers mix of small-town nostalgia, soapy family drama and high-tech sfx, this Dreary Android Runaway Yarn Lags way behind the Spielberg thoroughbreds it tries so hard to ape.' – Time Out

Verdict: Child-friendly robo-antics.

JOHNNY 5

Also known as: Number 5/ Strategic Artificially Intelligent Nuclear Transport/ SAINT No. 5.

Key Narratives: *Short Circuit* (film, 1986).

Appearance: A body made up of jointed pistons like a cartoon terminator, running on caterpillar tracks and sporting an oblong head with camera-lens eyes. Hi-tech but loveable.

Design classic? Probably looks a little messy by today's standards.

Origins: Prototype Cold War robot for US Military, developed by Dr. Newton Crosby (Steve Guttenberg).

Designation: Robot, for nuclear transport. Later sentient.

Weaponry: n/a

Fear Factor: 👽

Cuteness Factor: ♥♥♥

Artificial Intelligence: High.

Skills: Programmed for military use, but comes to reject this and to respect life.

Notability: ★★

Databank
- When military robot Number 5 disappears, he finds himself at the home of the animal-loving Stephanie (played by Ally Sheedy), who takes it upon herself to explain a few philosophical truths to him. We're on the side of Ally and the robot as he is hunted down and 'destroyed' by the military, but all they actually manage to destroy is a fake, made up of spare parts. Thank goodness.
- The film's concepts are based on various ideas and inspirations the creators had, including the personalities of the *Star Wars* droids and an educational video they made about what it would be like to have a robot living around the house doing everyday jobs.
- The film and its robot protagonist were originally going to be a lot darker. This film was originally conceived as a dark high-tech thriller. Number 5 was going to be a Terminator figure, a villainous, out-of-control, heavily-armed military robot which had escaped from the lab that had developed him. The hi-tech comedy it ended up as only emerged after several drafts…
- Johnny 5 is voiced by puppeteer and voice actor Tim Blaney (born 1959) who also provided voices in the *Men in Black* films, as well as various projects for Muppet creator Jim Henson.
- The robot was designed and built by Eric Allard of the All Effect Company, whose other notable creations include the Energizer Bunny and effects on the films *Demolition Man, Alien Resurrection* and *Stuart Little*.

Quotes
'*I'm Johnny 5, I'm alive!*'
'*Cuteness is never far off, though [director] Badham has enough sense of pace, and the robotics are sufficiently inventive, to keep the laughs coming.*' – Time Out

Further Appearances
The 1988 sequel *Short Circuit 2* – quite highly regarded as sequels go but without either of the original stars, Ally Sheedy or Steve Guttenberg. A script exists for a third *Short Circuit* movie written in 1989 and rewritten in 1990, but the project was scrapped. According to *Variety* magazine in April 2008, Dimension Films bought the rights to make a third *Short Circuit* movie, involving a boy from a broken family meeting and befriending Number 5. Whether this would be a sequel or a remake is not clear.

Assessment
There is a cluster of films in the 1980s exploring the extravagant adventures of teenagers caught up with the new worlds of computing, robotics and technology. It was the decade where the teen romance flick properly caught up with the sci-fi concerns of the 1950s/60s B-movie. We see a boy almost causing a nuclear holocaust from his bedroom in *War Games*, exploration of the then-new landscapes of virtual

reality in *Tron*, sentient computer love-triangle in *Electric Dreams* and the 'perfect virtual woman' created in *Weird Science*. Teenage dreams made real by the provision of technology. Here we see another variation on the theme – teen comedy meets 'robot gone bad'. An accidental bolt of lightning bringing to life another childhood fantasy – the toy robot which can think for itself. (Number 5 may be a tool of the military, but to all intents and purposes he is a very expensive toy. He even looks like one, although there's no official action figure on the market.) Rather than a cold RoboCop or a rampaging Terminator, though, Johnny 5 is an amenable chap who is only after what he calls 'imput' [sic]. Although the script ultimately aims for cuteness rather than menace, Johnny 5 still manages to convince as a potential 'working' robot, largely thanks to the clever design with the jointed arms, gripping hands, etc., and the emphasis of function over sleekness. He's still fondly remembered today.

Verdict: Almost alive…

MAX

Also known as: Artificial intelligence/pilot of Trimaxian Drone Ship.

Key Narratives: The 1986 children's sci-fi film *Flight of the Navigator*, written by Michael Burton and Matt MacManus, and directed by Randal Kleiser.

Appearance: Metallic probe-like extrusion with a large yellow 'eye' like a metal sunflower, represented the interface with the ship.

Design Classic? Rather beautiful, for the time. Very dark and metallic, in the manner of the 80s, but oddly timeless.

Origins: Originated on Phaleon and crashed on Earth. Max's mission is to collect data from other planets and take it back home. A boy called David was abducted by Max as a sample in 1978 and returned in 1986 without having aged. David is implanted with information which Max needs.

Designation: Artificial intelligence/computerised pilot.

Fear Factor: 👽

Cuteness Factor: ♥

Artificial Intelligence: High.

Skills: Faster-than-light drive, time-travel.

Notability: ✯✯✯

Databank
- One of the models used for Max's ship has been displayed since 1989 in the Disney-MGM Studios exhibition in Orlando, Florida.
- Director Randal Kleiser started his career in TV with shows including *Starsky and Hutch* (referenced in *Flight of the Navigator* when David asks if it's still on), and then went into movies, his other most famous assignment being *Grease*.
- The special effects supervisor was Jeffrey Kleiser, brother of director Randal, who went on to work on the *X-Men* films.

Quotes
'How many times have you done this?'
'Zero.'
'Zero! I'm not going to let you try this out on me! What if you fry my brain?!'

The *New York Times* was impressed, calling the film 'absorbing' and saying that it never talks down to its audience.

Memorable Moment
There's an amusing scene after Max absorbs information from David's brain and his cool, calm, robotic persona turns into a more exuberant, human-like one and the two begin trading insults…

Assessment
'Though its script and direction deserve highest praise, a great sense of humor helps to distinguish *Flight of the Navigator* from other strong science fiction films of the late 1970s and early 1980s. While its time-travel tale is a serious one, it deftly uses comedy to enhance the adventure and crank up the entertainment value. In many of the best ways, the film calls to mind the perfect blend of *Back to the Future*, without feeling the least bit derivative.' – *DVDizzy.com*

Verdict: Expertly navigated.

THE TRIPODS

Also known as: The Masters.

Key Narratives: The 1967-8 trilogy of novels by 'John Christopher' (one of the many pen-names for the SF writer Samuel Youd), *The White Mountains*, *The City of Gold and Lead* and *The Pool of Fire*, which were adapted into the BBC's teatime serial *The Tripods*. At least, the first two were. The plug was pulled on the series after two years before they could adapt Book 3, even though Series 2 ended on a cliffhanger. In 1988, Youd produced a prequel, *When the Tripods Came*.

Appearance: A bit of a cheat, this one, because the gigantic, three-legged Tripods, masters of a future Earth, at first appear to be robotic. In fact, the aliens – strongly reminiscent of H.G. Wells's Martians – are only robotic on the outside. The Tripods are travel vessels for the alien Masters to get around in.

Design classic? Still have a haunting beauty.

Origins: Outer space. It is revealed only in the prequel that the aliens managed to hypnotise Humanity by means of a TV show called *The Trippy Show*, which should give people cause to worry about *Britain's Got Talent* and *The X-Factor*.

Designation: Hostile, but appearing benevolent for their own ends.

Weaponry: Large, loud, colourful, scary lasers.

Fear Factor: 👽 👽 👽

Cuteness Factor: Not really.

Notability: ★★★

Databank
- Humanity has reverted to a pre-industrial age, living largely in villages, and ruled over by the alien masters, the great gleaming Tripods. The series has one of the best openings ever for a kids' teatime TV show, featuring some beautiful lighting and direction and, for the time, very impressive special effects which really give a sense of the scale of the first Tripod we see as it steps over a lake, its legs reflected in the water. The story opens with a 'capping' – the ceremony which all 14-year-olds must undergo – the placing of a metallic control 'cap' on the head to control all impulses of resistance and keep Humanity docile.
- The story focuses on three boys, Will, Henry and Beanpole, and their escape from the Tripods to find the 'free men' living in the White Mountains. Once they have found this community, they infiltrate the Tripod city, wearing fake 'caps', and the second book/series concentrates on this mission. In the third book, the resistance attacks and overcomes the Tripods' three cities by introducing alcohol into the water supply, followed by a bombardment via hot-air balloon.
- Everything augured well for *The Tripods*. It had a comparatively big budget and was jointly produced by the BBC and Australia's Seven network – a bold move at the time. The BBC supported it vocally, and it was even mooted as a replacement for the then-floundering *Doctor Who* – director Christopher Barry claimed *The Tripods* was a series that would 'run and run'. This seemed unlikely at the time, one had to admit, for a series based on a contained narrative of three novels – one whose first season seemed remarkably stretched and padded into 13 episodes with endless scenes of travelogue and grape-treading (yes, really), and some episodes in which the eponymous Tripods do not appear at all.

- Despite being recorded entirely on videotape – usually the stock medium of the cheap-and-cheerful (sitcoms and soaps) – the series manages to look very visually impressive, with lots of convincing special effects and experimental moments that actually work. A combination of models, state-of-the-art 80s computer graphics and other effects means that one is rarely pointing and jeering.
- By the time the lavish second series came along, with its impressive computerised rendition of the Masters' city, its harder sci-fi edge, and some of the questions of the first 13 episodes beginning to be answered, a lot of people had stopped caring. It's fair to say that a good few fans of the books were a tad miffed, though, at some of the liberties taken in the adaptation – not least the unlikely addition of a bar called The Pink Parrot where slaves go to relax during their 'time off'.

Quotes

'Although The Tripods *is very much a product of its time (especially the second series – more of that later) that does not mean it's any less of an essential watch. Most "classics" are "of their time" and whilst the series reeks of the Eighties, there's a charm and a warmth present that will keep you enthralled throughout all twenty-five episodes.'* – Den of Geek

Memorable Moment

One of the best moments in an otherwise slow first series comes when our young heroes are cornered by the Tripods in a forest at night and we get to see the three-legged beasties lit up in the darkness – actually looking genuinely menacing for the first time since their first appearance – and shooting bright bolts of fire into the night.

Assessment

Despite the dodgy moments in the adaptation, and the terrible padding in the first series, the Tripods impress as genuinely scary aliens – with the whole are they/aren't they robotic/cyborg creatures question giving an added frisson to the first 13 episodes. When the Tripods can be bothered to report for duty, that is. It remains a great shame that the series conclusion was not filmed, and the Tripods themselves remain memorable icons of 1980s TV.

Verdict: Two legs out of three isn't bad.

ROBOCOP

Also known as: OCP (Omni Consumer Products) Crime Prevention Unit 001. Formerly police officer Alex James Murphy.

Key Narratives: The film *RoboCop* (1987) and its sequels.

Appearance: Tall, human-shaped, muscular. Features hidden behind a close-fitting metallic helmet and visor. Mouth can be seen.

Design classic? The helmet and physique is instantly recognisable.

Origins: Murphy, killed in the line of duty, is turned into RoboCop by OCP's team, led by chief technician Bob Morton.

Designation: Cyborg police officer in Detroit.

Weaponry: Various, including the Auto-9 (a 9mm handgun), a 'gun arm' on the left arm which doubles as a rocket-launcher, and a 'data spike' in the right hand which can download information (used as a weapon even though not intended as one).

Fear Factor: 👽 👽 👽

Cuteness Factor: Difficult to love, really.

Artificial Intelligence: Controlled by his creators and by four 'prime directive' laws – 'Serve the public trust', 'Protect the innocent', 'Uphold the law' and a fourth, classified law, which turns out to be 'any attempt to arrest a senior OCP employee results in shutdown'. RoboCop malfunctions when he begins to remember aspects of his human life, and goes on a personal quest for vengeance, seeking out the gang who killed him as Murphy.

Skills: Crime prevention in a badass-cyborg kind of way.

Notability: ★★★

Databank
- His vision has an 'internal zoom' and several modes including thermal vision.
- He has audio-visual playback of everything he sees, so he can look over scenes again with total neutrality, unbiased by memory.
- Murphy's limbs are totally replaced with prosthetic ones, but his human nervous system is retained.
- RoboCop's armoured shell is made of titanium and kevlar, and is strongly resistant to heat. His visor conceals most of the face, which is Murphy's face but grafted on to a new, artificial skull.
- Like any machine, he still requires 'servicing' from time to time, and his organic systems need to be monitored. He is programmable, too, so he is largely under human control.
- Among the scenes storyboarded but never recorded was one where RoboCop visits his grave.
- Actor Peter Weller took the part of RoboCop for various reasons. He was a fan of director Paul Verhoeven, and also was intrigued by the 'medieval' themes of the script. 'I read the script and knew right away that it was my kind of film – its themes are massive,' he said in an interview in 2002. He was attracted by the idea of 'a corporate, mercantile society crushing everything in its wake, just like in the

Renaissance... Free trade and the massive flow of information have liberated society, but at the same time, you see an immense amount of greed... RoboCop is an allegory about imperialisation, technology and humanity – this society takes the life of this guy and also robs him of his innocence.'
- Just as the actors in the bodies of the Tin Man and Gort had done decades before, Weller found the cyborg costume so uncomfortable to wear that he could only remain in it for limited periods of time. He lost a frightening amount of weight through water loss, until an air-conditioning unit was installed in the costume. It took 11 hours to fit him into the suit.
- Weller learned some mime exercises with a mime coach to help him to express himself better through his movements, but found the suit too restricting to use what he had been taught. This caused a delay in filming.
- Some reviewers have seen RoboCop as a Christ figure, citing as evidence the obvious 'resurrection' and 'saviour' themes, and the fact he is seen walking on water during the movie's final showdown. Interviews with director Paul Verhoeven have indicated that he supports this reading of the film.

Quotes
Roger Ebert's review:

'Considering that he spends much of the movie hidden behind one kind of makeup device or another, [Peter] Weller does an impressive job of creating sympathy for his character. He is more "human," indeed, when he is a robocop than earlier in the movie, when he's an ordinary human being. His plight is appealing, and Nancy Allen is effective as the determined partner who wants to find out what really happened to him.'

'...a comic book movie that's definitely not for kids. The welding of extreme violence with four-letter words is tempered with gut-level humor and technical wizardry... Nancy Allen as Weller's partner (before he died) provides the only warmth in the film, wanting and encouraging RoboCop to listen to some of the human spirit that survived inside him. RoboCop is as tightly worked as a film can be, not a moment or line wasted.' – Variety

Further Appearances
RoboCop comic book series, animated TV series, live-action TV series and various video games.

Assessment
Where the other fearsome 80s icon the Terminator is all about remorseless, robotic perfection and unceasing pursuit, RoboCop, while looking equally fearsome, is a more tortured soul. His programming is not perfect, and in him we once again see the central preoccupation of science-fiction, especially cyborg fiction – what happens when man meets machine, when flesh is melded with technology? What are the true

implications of this? And, even with our bodies in a metal shell, do we ever lose what it really means to be human?

Verdict: He is the Law, but he's a bit of a tortured soul of a cyborg saviour.

DATA

Also known as: Lieutenant-Commander Data.

Key Narratives: *Star Trek: the Next Generation* TV series from 1987 onwards, and *Star Trek* films from *Generations* onwards.

Appearance: Humanoid, but with a pale complexion, slicked-back hair and yellow eyes.

Origins: Built by Dr Noonien Soong, a cyberneticist, who also built androids Lore and B-4.

Designation: Android, non-hostile, capable of self-defence.

Weaponry: Immense brain.

Fear Factor: 👽 👽

Cuteness Factor: ♥♥♥

Artificial Intelligence: Extremely High.

Skills: Recall and analysis. Data can be relied upon to provide a calm head in a crisis, and to analyse the situation without letting emotions get in the way.

Moral guidance: Data is driven by wanting to learn human emotions, and this notion forms the basis of many of the programme's Data-centric episodes. He is basically 'good', helpful, respectful, does no harm to others, and is a loyal member of the *Enterprise* crew.

Notability: ★★★★★

Databank
- When the first rumours came through of a relaunched *Star Trek* for the 1980s which would boldly go where the original series had never been, the puzzled reaction of many casual viewers was, 'What, so there's no Spock in it, then?' But it soon became clear that the new *Enterprise*'s mild-mannered android was going to be the spiritual successor to the original show's cool and calculating Vulcan. Like Spock, he was to offer calm advice to the Captain in the face of adversity, while learning about the quirky ways of humanity and its emotions. Data was a Tin Man for the cable TV age, an endearing character whose quest to understand the complexity of the human heart was usually touching rather than mawkish.

- The plan by *Star Trek* creator Gene Roddenberry, and the challenge for the writers and actor Brent Spiner, was for Data to become more and more 'human' as the show progressed.
- In another human touch, Data owned a cat called Spot which seemed to change gender during the series.
- Data also experiences anger (in the story 'Descent') and laughter for the first time, as a gift from superbeing Q in the episode 'Deja Q'.
- Data's 'evil twin' brother Lore first turns up in 'Datalore', in which he is reassembled, in 'Brothers', in which he poses as Data and is fitted with an emotion chip, and then in 'Descent', a two-part story with a central cliffhanger bridging the sixth and seventh series, in which Lore is in league with the Borg. Lore's story is brought to an end when he is permanently deactivated and the emotion chip removed.
- Data finally installs the emotion chip in himself in the *Generations* film, and experiences the full range of human emotions for the first time.
- As the *Next Generation* characters' stories are brought to a close, Data dies sacrificing himself to save Captain Picard in the 2002 film *Nemesis*.

Quotes

'Feelings aren't positive and negative. They simply exist. It's what we do with those feelings that becomes good or bad.' – Data, in the episode 'Descent'.

'I don't see no points in your ears, boy. But you sound like a Vulcan.'
'No, sir. I am an android.'
'Almost as bad.'

– Exchange between Admiral McCoy and Data in the first ever *Next Generation* episode, 'Encounter at Farpoint'.

Memorable Moment

A key point in Data's story is the episode 'The Measure of a Man', from the second *Next Generation* series first shown in 1989. In this, Data must avoid an 'experimental refit' by a cyberneticist by resigning from Starfleet. In order to do this, he must submit to a hearing to determine that he is a Starfleet officer, capable of resigning, and not simply a piece of property. In something of a landmark for the series, Data is ruled to be a sentient being.

Assessment

Data proved to be a hugely popular character with both Trekkies and the general viewing public, thanks largely to the scripting and the work of actor Brent Spiner, who managed to make his vulnerability endearing rather than mawkish. He is the personification of *Star Trek*'s optimistic, humanist, utopian outlook. If all sci-fi can be said, on some level, to be about humanity's relationship with technology, then

Data embodies the programme's largely positive view of scientific development and its benefits to humanity. He is pivotal to a number of episodes, and it could be argued that he goes on the greatest character journey of any of the regulars over the course of 178 episodes and four films. The bold decision to kill him off was not taken lightly, and will ensure that his popularity will remain undimmed.

Verdict: Logical and loveable.

THE BORG

Also known as: The Borg Collective.

Key Narratives: The TV series *Star Trek: The Next Generation*, *Star Trek: Deep Space Nine* and *Star Trek: Voyager*, and the film *Star Trek: First Contact*. The Borg replaced the Klingons in the 1990s as the *Star Trek* universe's most fearsome foes.

Appearance: Cyborg zombies. Terrifyingly human-like, but with blank, pale faces, and surrounded by complex cybernetics, like living suits of circuitry and electrical equipment, whose extrusions and probes penetrate their unnaturally white skin. In *First Contact* we get to see their Queen, a strangely compelling, pale creature played by the strikingly beautiful actress Alice Krige.

Design classic? Beautifully captures the body-horror of the cyborg experience.

Origins: Unclear from TV episodes or films, although we know they began as an organic race thousands of years before. However, the graphic novel *Star Trek: The Manga* suggests they are a malfunctioning experiment in nanotechnology.

Designation: Cyborg collective – hostile.

Weaponry/skills: 'Assimilation'. Their 'cube' and 'sphere' ships are also extremely powerful and dangerous in space warfare.

Fear Factor: 👽👽👽👽👽 (New *Star Trek* needed a frightening arch-adversary and got one in spades.)

Cuteness Factor: ♥♥ (But that's just for Seven of Nine.)

Artificial Intelligence: High and collective.

Catchphrases: 'Resistance is futile', perhaps a play on the old standby 'resistance is useless.'

Notability: ✯✯✯✯✯

Databank
- Each Borg 'unit' has an artificial eye replacing the organic one, which enables it to see beyond the normal spectrum that humans can see. Its body is connected to

various different cybernetic devices. Rather than replacing the organic, as in the original idea for the Cybermen, the cybernetic devices work with the original organic body.
- Each Borg has a 'cortical node' controlling every other implanted 'fixed location' cybernetic device within a Borg's body. It is usually implanted in the forehead above the organic right eye. If the cortical node fails, the afflicted individual Borg eventually dies, as it cannot be replicated or repaired. Successful replacement of the node can be carried out on board a Borg vessel, but only if the failure is detected promptly before the Borg's impending death.
- In the episode 'I, Borg', an injured Borg drone called 'Third of Five' is discovered in a damaged scout ship. He is given the name 'Hugh' by the *Enterprise*'s chief engineer Geordi LaForge, and begins to understand the concept of individuality. In the later episode 'Descent', Hugh is seen as part of (and later leader of) a group of rogue or 'rebel' Borg, who have 'assimilated' the idea of individuality via Hugh. These are controlled by Data's evil brother Lore.
- *Star Trek: Voyager*'s regular crew features the character Seven Of Nine (Tertiary Adjunct of Unimatrix Zero One), played by Jeri Ryan, who was assimilated by the Borg as a child but whose humanity had started to reassert itself. As a human she was called Annika Hansen. She has no formal link to the Collective but can sense other Borg.
- The initial concept of the Borg was for them to resemble insects, from which the 'hive mind' idea emerged. Although this visualisation of them was abandoned as being too expensive to realise, the 'hive' idea remained.

Quotes

'We are the Borg. Existence, as you know it, is over. We will add your biological and technological distinctiveness to our own. Resistance is futile.'

Memorable Moment

In the classic two-part episode 'The Best of Both Worlds', Jean-Luc Picard is captured, surgically altered, and turned into 'Locutus of Borg' – a bridge between humanity and the Borg. After the Borg are defeated, Dr Crusher and Data remove all Borg implants from Picard, but the experience is seen to remain with him for years afterwards.

Assessment

If the Cybermen represented their creators' concerns over cybernetics in the decade of the first human heart transplant and advances in prosthetics, then the Borg reflect the 1990s obsessions with the interaction between the organic and the mechanical at a very close level. As terms like 'cyberspace' and 'wetware' became common currency in science fiction, the Borg capture the zeitgeist with their chillingly effective embodiment of the fusion between man and machine, a 'cybernisation' process which not only removes emotion but takes it a step further – it removes individual will and subverts the ego to the common cause. The Borg are frightening on several levels: visually, thanks to their body-horror appearance; in a narrative sense because of their apparent unstoppability; and also in the intellectual sphere because of what they represent, the removal of what actually makes us human – free will. To see Captain Picard, the admirable, humanist, crusading, firm-but-fair Starfleet officer who the viewers have come to know and trust over several series, subverted to their cause is not only visually shocking but chilling to the core in its implications. It's a credit to the writers that the 'Borg experience' is seen to stay with Picard, haunting him for years afterwards.

Verdict: Collectively terrifying.

Chapter 12

Living Ships

Part robot, part organic, part space vessel – the ships which have a special relationship with their owners and just seem to know what to do without being told…

The TARDIS from *Doctor Who* (1963 onwards)
The Doctor anthropomorphises **the TARDIS** on numerous occasions, calling his time-ship 'she' and 'old girl'. It's often hinted that they exist in some kind of symbiotic, telepathic relationship, and the Doctor's attempts to get the TARDIS to do what he wants often involve coaxing, coercion and gentle stroking of the console. And sometimes, mainly from Tom Baker's Doctor, it took a good thump – which by the time Christopher Eccleston took over in 2005 had become a hearty whack with a hammer. The TARDIS, it is continually implied, is sentient – and, at times, as cantankerous, mercurial and unpredictable as the Time Lord himself. In the 2011 episode 'The Doctor's Wife', we see the first manifestation of the TARDIS in speaking form, as a lissom, rather Gothic-looking female called Idris. Better than the horrific movie rumour which was going round in the 1990s, about the console being fitted with a giant pair of rapping lips…

The Liberator from TV's *Blake's 7* (1978-81)
By the end of the second episode of Terry Nation's space opera, the interplanetary rebels have escaped their prison ship and have gained control of a beautiful, cathedralesque spaceship called the **Liberator**. There are certainly hints that **Zen**, the impassive computer interface working for Roj Blake and his friends, 'is' the Liberator in a sense. It controls all the ship's defence and attack mechanisms and seems to be in charge of stuff like when to dim the lights, and so on… When the crew first encounter the ship, it attacks them with psychic projections to scare them off, and it's capable of making the odd, tubular guns red-hot to the touch so that only one person can use each weapon. Built by the mysterious System, it is destroyed after it flies through a cloud of corrosive micro-particles. The crew's stand-in ship Scorpio, a battered freighter with a few whizz-bang adaptations, gets them around in the fourth series, and is nowhere near as impressive or beautiful.

Max (from *Flight of the Navigator*, 1986)
A ship in need of important information from a small boy's mind, which rather disturbingly abducts him in 1978 and returns him, the same age, in 1986. See separate entry.

The Vorlon and Shadow ships (*Babylon 5*, 1993-99)
Writer and executive producer J. Michael Straczynski once stated that one of the edicts for his epic 'novel for TV' was that there would be 'no cute kids, no cute robots'. And indeed there aren't, but we do see some rather scary examples of part-organic technology. The dark, manipulative **Shadows** and the supposedly benevolent, godlike **Vorlons** are two ancient races who have been pulling Humanity's strings for millennia, forcing the human race to evolve through conflict and survival. The symbiotic Shadow ships, each with an organic pilot, are masterfully frightening – dark, glistening, spider-like monstrosities which move with a 'screaming' noise and cut through spaceships with deadly 'slicing' rays. The greenish, flower-like Vorlon spacecraft can sing and are seen to extrude appendages.

The Lexx (from *Lexx*, 1997-2002)
Not only a sentient spaceship, but also a powerful weapon with the ability to destroy entire planets – and enjoy it too. With a shape like the head of a giant insect, **the Lexx** is a 'bioship', controlled by living energy known as 'the Key'. It is inclined to take orders literally, and becomes pregnant thanks to a dragonfly. It speaks with a male voice, although the voice used for the dubbed-into-German version is female… . The series is unusual in that the story unfolds over thousands of years (some of which the crew spends in cryogenic stasis), and also for some sexual themes rivalled only by its contemporary *Farscape*. In the third season, the plot is driven by Lexx running out of food and having to fly slowly to conserve energy, and in the fourth they finally find the dangerous planet of Earth at the heart of the 'Dark Universe'…

Moya (*Farscape*, 1999-2003)
The enjoyable Australian-American TV series *Farscape* brought a wealth of wit, inventiveness and wacky characters to the screen, and such was its zest and verve that it didn't seem to matter that a lot of its best ideas were reminiscent at times of *Blake's 7* and *Babylon 5*. As well as being more charged and heady with sexual charisma than any of its sci-fi cousins, *Farscape* was never afraid to explore the old sci-fi idea of the relationship between the organic and the machine. Superficially similar to Lexx, **Moya, the Leviathan ship**, in which the rag-tag crew is on the run from the Peacekeepers, is alive. It's a peaceful creature with no weapons capabilities, and the crew tries to be sensitive to her needs and wishes. She exists in a symbiotic relationship with the blue-skinned alien **Pilot**, whose 'steering' of her

appears driven by coaxing and persuasion, as much as by skill and technology. It's often Pilot's job to tell the crew what Moya will and won't do in a dangerous situation. Moya's day-to-day well-being is looked after by the **DRDs**, small, scuttly, egg-shaped repair-drones with glowing antennae. As well as repairs, they can locate lost items and have some defence capabilities. And things become even more complicated when it turns out the spaceship is actually pregnant… Talyn, Moya's offspring, turns out to be a hybrid of Leviathan and Peacekeeper technology, and packs a few guns away too.

Andromeda Ascendant (Gene Roddenberry's *Andromeda*, 2000-05)
A ship dating from the time of the old 'Systems Commonwealth', the Andromeda's controlling Artificial Intelligence projects its artificial personality in the form of the lovely Lexa Doig. It's implied that there's a strong emotional attachment to Captain Dylan Hunt, too. James T. Kirk and the *Enterprise* may have had their moments, but they didn't manage to get quite so close. Its complete official designation is 'Shining Path to Truth and Knowledge AI model GRA 112, serial no. XMC-10284.'

Chapter 13

The Modern Age

KRYTEN

Also known as: Kryten 2X4B 523P, to give him his full designation.

Key Narratives: The BBC TV sci-fi comedy series *Red Dwarf* which ran for nine series from 1988 (a revival series *Red Dwarf: Back To Earth* was shown on the Dave channel in 2009).

Appearance: Vaguely humanoid, but with a squared-off, lumpy face.

Origins: Built by the corporation DivaDroid International in 2340; one of a number of Series 4000 models based on a design by roboticist Professor Mamet.

Designation: Slave mechanoid.

Fear Factor: 👽

Cuteness Factor: ♥

Artificial Intelligence: High.

Skills: Science expertise, and, er, winding up Arnold Rimmer…

Notability: ★★★

Databank

- The late 1980s were not a good time for science fiction on the BBC. *Doctor Who* was struggling, ratings-wise if not creatively; the *Blake's Seven* crew showed no sign of a resurrection following their apparent massacre in 1981; the lavish children's adventure *The Tripods*, planned as a trilogy, was canned after two series; and the innovative and well-made *Star Cops* was put out in the summer on BBC2 with no publicity, and died a death after just one nine-part series. It seemed audiences wanted soap grimness and endless police and fire-fighter dramas. However, a little science-fiction show found a home on BBC2 and a loyal following simply by being smuggled out under the blanket of comedy – Rob Grant and Doug Naylor's *Red Dwarf*.

- Although played for laughs in terms of the situation and dialogue, the ideas in *Red Dwarf* are no more alarming or 'silly' than those found in conventional 'straight' sci-fi – a giant ship left deserted apart from one crew-member, accompanied by a hologram of his arch-enemy, a being evolved from the ship's cat and a sentient computer.
- Although the humour is at times quite broad and down-to-earth – mostly thanks to Craig Charles's portrayal of hapless 'last human' Dave Lister – the concepts are, at times, very clever and mind-bending, with episodes about body-swapping, memory exchange and virtual reality. The feel is also not a million miles away from that of Douglas Adams's *The Hitch-Hiker's Guide To The Galaxy*, with a ragtag bunch of flawed but basically mostly likeable heroes roaming the galaxy and trying to survive in a variety of increasingly bizarre situations. Into this gang in the second series comes Kryten, the robot, who was to become a series regular.
- Kryten was initially a one-off character, a robot butler, his name a pun on the 'Admirable Crichton', the resourceful butler in the J.M. Barrie play and subsequent film. In his first Series 2 appearance he was played by David Ross.
- The character re-appears at the start of the third series, now looking (and sounding) a bit different, and played by Robert Llewellyn.
- Kryten's accent was a subject of much discussion between Llewellyn and the crew. He eventually went for something which was supposed to be Canadian, an experiment which the actor has admitted in recent years did not quite come off.
- Like *Star Trek*'s Data, Kryten longs to be human. In one episode, the gang discover a DNA modifier and Kryten does temporarily become a human being.
- Lister shows Kryten classic anti-authority films *The Wild One*, *Easy Rider* and *Rebel Without a Cause*, after which the robot responds to Rimmer's 'What are you rebelling against?' with the dialogue from *The Wild One*, 'Whaddya got?'

Further Appearances
Red Dwarf USA, a pilot made in 1992 but never broadcast. Robert Llewellyn was the only member of the original cast to reprise his role.

Memorable Moment
It's hard to forget Kryten's first appearance on the distressed spaceship Nova-5. Calling Red Dwarf on a video-link, he enlists the help of Lister and friends by displaying a number of attractive female crew-members on their monitor. When we get there, we see that the ladies in question are in a less than healthy state. In fact, they have been dead for centuries and are skeletons. Hologrammatic officer Arnold Rimmer is prompted to describe Kryten as 'the android version of Norman Bates'.

Assessment
In the manic atmosphere of *Red Dwarf*, where the crew seem to survive and to look out for each other despite their apparent mutual hatred, Kryten is an interesting addition to the mix, and often has some of the best lines.

Verdict: Admirable robot.

DR. ARLISS LOVELESS

Key Narratives: The 1999 'steampunk' movie *Wild Wild West*.

Appearance: Pointy-bearded human in a steam-powered wheelchair who has lost the lower half of his body in the Confederate War. The chair sometimes turns into a mechanical spider on hydraulic legs.

Designation: Villainous cyborg.

Weaponry: Various, including his newest, a steam-powered tank.

Fear Factor: 👽 👽

Cuteness Factor: Err…

Artificial Intelligence: High.

Skills: Invention and Machiavellian cunning.

Notability: ✯✯

Databank

- Played by Kenneth Branagh, who, it's fair to say, probably doesn't consider the role to the pinnacle of a varied and critically-acclaimed career.
- Branagh suffered for his art. When he was in Loveless's articulated metal platform, he had to be seated in the device in a kneeling position. He needed to get up every few minutes and walk around to get the circulation back in his legs, as they would keep going numb from being in that position for an extended period.
- Based on the character of Dr. Miguelito Quixote Loveless in the 1960s TV series *The Wild Wild West*, a brilliant, insane dwarf.
- His villainous plan is to destroy the USA with his mechanised troops, unless President Grant divides them up between various European countries, Mexico and Loveless himself.
- In the final battle with the hero James West, Loveless plunges to his doom over the edge of a canyon.
- The film didn't go down that well with critics, and won several of that year's 'Razzies' for the worst offences in cinema. Robert Conrad, star of the original 1960s TV show, made his dislike of the movie very public. The star Will Smith is said to have been embarrassed by the film's success – it earned almost $50 million in its opening weekend. Years later, Smith apologised publicly to Robert Conrad, saying he understood his anger and criticism of the film version, as well as Conrad's refusal to make a cameo appearance in it.

Quotes

'We may have lost the war, but we haven't lost our sense of humor. Even when we lose a lung, a spleen, a bladder, thirty-five feet of small intestine, two legs, and our ability to reproduce – all in the name of the south – do we ever lose our sense of humour?' – Dr Loveless

'The diminutive villain Loveless of the TV series now has lost the lower half of his body, which makes him literally loveless. He and West trade comic insults that go for the jugular. Branagh bristles with glinty-eyed fake Southern cheer. Not only is he bitter about the loss of his appendages, but he is also an unreconstructed Southerner who has never forgiven the surrender at *Appomattox*.' – San Francisco Chronicle

Assessment

A memorable villain from a less-than-memorable film, but Dr Loveless probably won't be fondly remembered by fans of either Kenneth Branagh or of the original series.

Verdict: Gone west.

SIR KILLALOT

Key Narratives: The TV game show *Robot Wars*, 1998 to 2004.

Appearance: Fearsome-looking beast, like a cross between a small Dalek, a giant cockroach and a squat, heavily-armoured knight.

Design classic? Bristling with weaponry. Still looks tough today.

Origins: Built by the BBC's Special Effects department as the most fearsome of the 'House Robots' which put competitors to the test, and often considered the 'leader'.

Designation: House Robot – fearsome.

Weaponry: Lance, hydraulic cutter, flamethrower.

Fear Factor: 👽 👽 👽

Cuteness Factor: ♥

Artificial Intelligence: Apparently quite high.

Skills: Cunning, battle skills, attack, defence and tenacity.

Notability: ★★★

Databank

- *Robot Wars*, hugely loved throughout its run on the BBC and then on Five, was an inspired concept – a teatime game-show which was basically *Top Gear* meets

Star Wars, with a dash of crazy Heinz Wolff inventor show *The Great Egg Race.* The first series even had *Top Gear*'s Jeremy Clarkson hosting in a long leather coat, trying to do his usual mix of deadpan gravitas and schoolboy enthusiasm. He was replaced after the first series with Craig Charles from *Red Dwarf,* who managed to give the whole thing a more manic edge.

- Each week, robots built by teams of competitors (usually university engineering departments) would come to do battle with each other in various cleverly-constructed live-action games, involving mazes, ramps, pits and other hazards. The mechanical wannabes had to contend also with the less-than-tender ministrations of the roving House Robots and their deadly appliances.
- Sir Killalot was king among them, but also popular were Sergeant Bash, with his flamethrower and ramming arm, and the iron lady Matilda, a triceratops-like creature with hydraulic tusks.
- Killalot weighed 200 kilos, was 120cm long, had a steel exo-skeleton and was driven by a 1 horsepower, 24-volt motor.
- His lance was very effective thanks to its size (two metres) and its ability to rotate. With this, Killalot could lift the hapless robots, manoeuvring or pinning them like an expert wrestler.
- His cutter could exert a force of 15 tons at the tip, and could cut through steel bars and even wheels.
- It's easy to see why the show was so popular among teenage boys and grown-up robo-geeks everywhere. It had a perfect combination of factors – the tension of the competition, whipped up into a frenzy by the whooping *Gladiators* -style audience in a giant arena, and the metal death and destruction as the combatants locked steel horns.
- A dash of glamour was added by the lovely Philippa Forrester, who would slip with flirty sylph-like charm among the shy, bearded, bespectacled competitors, thrusting a microphone under their noses as she showed off her splendidly lissom form in leather trousers. (Later series would see Julia Reed and Jayne Middlemiss performing Robo-babe duties.)

Memorable Moment
That terrible, but also oddly satisfying moment in most episodes of the show when Killalot would get some hapless robo-novice on the ropes and lay into it with the hydraulic cutter. Sparks flew, the audience gasped and cheered, and you could see the teams of boffins wincing and covering their eyes as the result of several months' careful work in the lab was sliced through like butter.

Assessment
Given the potential for *schadenfreude* in seeing a robot competitor go up in a blaze of sparks, it wasn't surprising that the audience often seemed to be on the side of the

fearsome, and supposedly villainous, House Robots. If Killalot had an Achilles heel at all, it was his lumbering speed (thanks to his great weight), which meant that some of the more nimble competitors could dodge out of his way. Also, both his weapons were forward-facing which left him vulnerable from the rear. But all in all, his weight, power and sheer lumbering presence made him a formidable opponent and one of TV's most memorable robots of the twenty-first century.

Verdict: A Dark Knight among robots.

BENDER

Also known as: Bender Bending Rodriguez, Bending Unit 22.

Key Narratives: The animated TV series *Futurama*, 1998-2003 and 2008 to present.

Appearance: Cylindrical body, domed head with bulging eyes, flexible limbs. Usually seen with cigar and bottle.

Design classic? Hard to say – he is intended to look comic, after all.

Origins: Manufactured by Mom's Friendly Robot Company. His mother was an industrial robot.

Designation: Robot, works on U.S.S. Planet Express Ship, as ship's cook, then Assistant Manager of Sales.

Weaponry: Various interesting components and objects inside his chest cavity.

Fear Factor: ♥

Cuteness Factor: ♥♥♥

Artificial Intelligence: High.

Skills: Strength. He is able to bend very hard materials. He is incredibly durable, almost indestructible, and is seen to survive being flattened. Surviving underwater. Video/audio recording. Disassembling and reassembling his robot body. Is shown to be able to lie, and can mimic human activities such as crying and belching.

Catchphrases: 'Bite my shiny metal ass', 'we're boned' and 'cheese it!'

Notability: ✯✯

Databank
- *Futurama* is created by Matt Groening, who brought us *The Simpsons*.
- Bender's extended family includes, among others: Professor Farnsworth, his owner and the creator of the sports utility robot on which Bender is based; his 'good twin' Flexo, another Bending Unit of the same model (now presumed dead, Flexo's

goatee-bearded appearance in the show is partly a satire on the *Star Trek* episode 'Mirror Mirror', in which the 'evil' Spock has a similar goatee); his father, who was killed by a giant can opener; Uncle Vladimir, who lived in Thermostadt, capital of the Robo-Hungarian Empire, and died at the age of 211; his cousin Tandy; plus Albert, Sally, Nina, Sam and eight other orphans, briefly adopted from the Cookieville Minimum-Security Orphanarium. His grandmother, a bulldozer, is also mentioned.

- Bender's head can be unscrewed, as first shown in the second episode 'The Series has Landed'.
- Bender survives being flattened, shot, melted, sliced in half and having a bomb exploded inside him… he would seem to be indestructible! The Terminator could learn a thing or two from him…
- However, he has no 'backup unit' into which his personality and memories can be downloaded so if his body is ever destroyed, it really is the end of him. This is established in the 2010 episode 'Lethal Inspection', in which Bender has to come to terms with his new-found mortality.
- What is Bender actually made of? Information on this is unreliable, and as Bender himself lies (or simply isn't bothered about whether what he is saying is true), conflicting percentages have emerged over the years. At times he has claimed that forty per cent of his body is zinc/titanium/dolomite/lead/chromium, and that thirty per cent of it is iron – while also claiming sixty per cent 'storage space'.

Assessment

Bender would seem in many ways to be the ultimate robot. He has a personality which enables him to interact with the human race and appears to have learned a thing or two about the deviousness required to survive among them, as well as experiencing human sensations of mortality. He even takes up religion ('Robotology') for a while, as seen in the episode 'Hell is Other Robots'. His body is, it appears, near-indestructible and he seems able to turn his hand to any number of tasks, as well as being equipped with a huge range of accessories. In the future, will we all want robots like Bender?

Verdict: Unswerving.

THE SENTINELS

Also known as: Squiddies, Calamari.

Key Narratives: The film *The Matrix* (2000) and its two sequels.

Appearance: Multi-tentacled, fierce, octopoid creatures with sharp claws.

Design classic? Well, they're unlike most other robots. One has to say 'yes.'

Origins: Created originally by humans for cheap labour, and turned nasty later…

Designation: Guard robots/ killing machines.

Weaponry: Clawed tentacles, all-seeing compound eyes, lasers, guided explosives called 'tow bombs' which can rip through a hovercraft… they are about as armed as they come! Also have audio sensors to detect the tiniest sound. If ever a robot had 'seriously don't want to mess with this one' written all over it…

Fear Factor: 👽 👽 👽

Cuteness Factor: Um, no.

Artificial Intelligence: Unknown.

Skills: Security, battle, combat, Search & Destroy missions.

Notability: ✸✸

Databank
- In 2199, Humanity is fighting a war against the machines, and what our hacker hero Neo initially believes to be reality is actually a giant virtual reality projection.
- The squid-like Sentinels patrol the sewers and passageways of the dead human cities.
- Their various appendages act as protection, batting off bullets and other projectiles to defend each other's 'head' units. Hence they tend to hunt in swarms or packs.
- The 'tow bomb' is a guided missile which resembles a small Sentinel. One Sentinel can release several at once when it is close enough to its target.
- The Sentinels can be killed by humans with a device known as a 'lightning rifle', a short-range weapon with a slow recharge. It emits an intense burst of lightning which can destroy a Sentinel.

Assessment
Fearsome and impressive on screen, and would have been difficult to realise before the CGI-propelled 1990s. The Sentinels are built up as being terrifying and indestructible, which of course is something of a two-edged sword for any robotic foe because at some point, our heroes are going to have to find a way to overpower them, or things could get very tedious. The 'lightning rifle' is a useful compromise for the writers – it's a get-out clause but not such an easy one that it looks like a cop-out. One thing's for sure – after seeing the Sentinels, you won't feel like eating squid again for a while.

Verdict: Sick squid.

WALL-E

Also known as: Waste Allocation Load Lifter – Earth Class (to give him his full title).

Key Narratives: The film *WALL-E* (2008).

Appearance: Metallic binocular-lenses-like head on a hinged 'arm' neck, surmounting a box-like body with a pair of caterpillar-tracks. Shovel-like hand appendages.

Design classic? Cute but efficient. Holds up well.

Origins: In the year 2805, WALL-E is one of the robots designated to clean up a polluted Earth.

Designation: Waste-cleaning robot – trash compactor.

Weaponry: n/a

Fear Factor: 👽

Cuteness Factor: ♥♥♥♥

Artificial Intelligence: WALL-E developing unexpected sentience is part of the story…

Skills: Ostensibly just cleaning and tidying…

Notability: ✯✯✯

Databank

- Mega-corporation Buy'n'Large has left twenty-ninth-century Earth in a polluted state, and so humanity was evacuated in starships while a team of robots cleaned up – supposedly for a period of five years. But the Earth became too polluted for the human race to return.
- WALL-E befriends the sleek, polished, white robot EVE, an Extraterrestrial Vegetation Evaluator, who arrives from the spaceship *Axiom*. Various twists and turns ensue in a tale of deception, vegetation and robot love… with a happy ending.
- WALL-E is the only robot of his kind still functioning on Earth. He's seen to collect spare parts for himself (which is lucky, as EVE turns out to need them later) and he replaces broken or worn out parts by cannibalizing other, dead WALL-Es. He's distracted from his work by collecting trinkets of interest and chatting to his cockroach friend Hal.
- Items collected by WALL-E include a cage full of rubber ducks, a Rubik's Cube and disposable cups filled with plastic cutlery. He is also seen finding a bra and a table-tennis bat and ball. He has an iPod with a magnifier for watching old

musicals. And a piece of debris seen falling away from WALL-E as he leaves Earth's atmosphere is the satellite Sputnik I, the first man-made object to be placed in orbit.
- Sound effects designer Ben Burtt, who also produced many of the memorable sounds for *Star Wars* (Artoo's burbling, the light-sabre hum) produced the electronic 'voice' of WALL-E.
- It's easy to infer that the film-makers were fans of the work of Steve Jobs, as the movie is chock-full of references to Apple computers. The film contains numerous references to Apple computers: the Apple boot-up sound as WALL-E is recharged, the use of the iPod, and EVE even looks a bit like a sleek iMac. Of course, it's no coincidence as this is a Pixar film, and Steve Jobs was CEO of Pixar until its acquisition by Disney in 2005, and was still a shareholder and member of the Disney Board of Directors when the film was made!
- Some critics didn't like what they saw as the 'left-wing' environmental subtext of the film. The English version of the online newspaper ohmynews.com claims the film 'paints in massive brush strokes, attempting to educate younger audience members with horrific vistas of a polluted, wasted Earth' and sees this as hypocritical because the makers are 'making sure Wall-E is endearing enough to use on games, toys, and stuffed animals so all concerned make a mint off of vulnerable family audiences.'

Quotes

'…this vision of an optimist surviving a pretty rough patch in his planet's history just plain works. Like Voltaire's Candide, WALL·E learns to tend to the garden. While I may argue with the little guy's taste in musicals, it's remarkable to see any film, in any genre, blend honest sentiment with genuine wit and a visual landscape unlike any other.' – Chicago Tribune

'…it's Pixar's big, bold belief that the mass audience will be astute enough to follow the visual clues and game enough to play along. So confident is the studio in its ability to charm audiences, it has made a futurist movie that's a lot like an old silent picture.' – Time Magazine:

Memorable Moment
EVE's first arrival, when she zips around in an elegant robotic aerial ballet before blasting a hole in the landscape with a flip of one robotic fin.

Assessment
It's hard to get the audience rooting for a piece of metal, but it works with Artoo Detoo and it works here… Sci-fi, humour and comedy come together in a likeable robot for all the family. It's hard to take seriously the hysterical idea that kids are being indoctrinated into becoming dangerous radicals through WALL-E's cuteness

and environmental nurturing. A redemptive but not-too-kitschy ending leaves things wide open for a sequel too.

Verdict: Not rubbish.

THE TESELECTA

Also known as: Justice Department Vehicle 6018.

Key Narratives: The *Doctor Who* episodes 'Let's Kill Hitler' and 'The Wedding of River Song', from 2011.

Appearance: Anything it likes, it would seem, although is mostly seen to take humanoid form, with a 'tessellating' shape-shifter effect. It is unclear if there are any limits on the size to which the Teselecta can adapt. Could it take the form of a child, for example, or a woolly mammoth?

Design classic? The jury is still out…

Origins: Constructed by the Justice Department.

Designation: Shape-shifting robot/android.

Weaponry/Skills: Morphing technology, surveillance, and apparently time-travel. Protected inside by the Antibody robots. Miniaturisation ray. Compression field to keep the crew miniaturised.

Fear Factor: ……

Cuteness Factor: ♥♥♥ (but only as Amy Pond).

Artificial Intelligence: None itself, but controlled by an intelligent and resourceful crew of 421 miniaturised humans, led by Captain Carter.

Notability: ★★

Databank

- The Teselecta locates notorious criminals near the end of their timeline and brings them to justice, aided by its shape-shifting technology which enables it to access places it otherwise could not (such as Hitler's office).
- It can be a little unwieldy and in need of having its balance adjusted/rectified when in a new form for the first time – causing the crew inside to be jolted about in their seats.
- Elevators inside allow the crew to access the android's more distant sectors, such as the eyeball.
- Forms we see the Teselecta take include Nazi officer Erich Zimmermann (it has to purloin his glasses, implying it can't replicate those), a Nazi soldier complete

with motorbike, the Doctor's companion Amy Pond, Gideon Vandaleur (former envoy of the Silence) and the Doctor himself, complete with false half-regeneration and death. (It takes 13 episodes for us to find out that it's faked.)
- The Teselecta must be pretty resilient. The Doctor's companions cremate what they think is his body, Viking-style, and the Doctor later remarks that he was 'barely even singed'.
- When fans complained that it was unlikely that the Teselecta could simulate Time Lord regeneration, *Doctor Who* writer Steven Moffat pointed out that the visual effects people of 2011 could do it, so 'you think the Teselecta can't?!' Fair enough, really. Nobody has ever accused alien/future technology of lacking the budget for convincing special effects before.

Nagging Questions

The Teselecta can obviously replicate clothing, and presumably headgear as we see it as 'The Doctor' wearing a stetson. (At this point we don't know the Doctor is fake.) A bullet from River Song's gun blows the stetson off the fake Doctor's head (she is just saying 'hello' in her inimitable way)… so what happens there? Does the Teselecta rapidly accommodate for this, or does it lose a chunk?

Memorable Moment

The revelation that the 'regenerating' and 'killed' Doctor was the Teselecta all along – when River Song looks into the eye of 'The Doctor' and sees him waving out. (She's in an alternative/possible/screwed-up timeline at the time – it's complicated.)

Assessment

Hugely entertaining when we first realise what's going on, the Teselecta is like a lighthearted cross between the Terminator and those *Beano* creations The Numskulls, tiny creatures who live inside and control a boy called Edd Case. It's also reminiscent of an amusing German YouTube video showing George W. Bush being started up and controlled by the little 'driver' who lives inside him.

Verdict: Is it a vehicle, a spaceship, an android? Ch-ch-changes the way we think about robots.

Chapter 14

Danger, Humans!

The Top Five Deadliest Robots and Cyborgs
Never mind protecting Humanity and not letting it come to harm, in the spirit of Isaac Asimov. What about those robots you really don't want to meet across a battlefield, down a dark alley or in a spaceship corridor? When it comes to deadly fire-power and general killing ability, these are the nastiest of the bad, mad metal machines. Our panel of experts has voted, the results have been sifted through Colossus and are being displayed on the LED screen…

Coming in at **No.5** it's **Gort** from *The Day the Earth Stood Still,* fearsomely implacable with his visor, deadly ray and earth-stopping powers. If he arrived today, he could enslave not only the world's aeroplanes and hydro-electric plants but also worm his way into the Internet, corrupting a trillion files and bringing the world's governments to their knees.

Our **No.4** is *Westworld*'s **Gunslinger**, dark and impassive and single-minded. Proving that there is nothing so terrifying as efficiency, this black-hatted killing machine even keeps going after he's been half-blinded and set on fire.

At **No.3** it's the *Star Trek* universe's **Borg Collective**. What could be nastier, and simpler, than a machine-based culture which just wants to 'assimilate' you, taking away all vestige of what makes you human? They don't only want to defeat the human race, they want to make sure there is no shred of humanity left inside us. They want us to become them.

It's back to straightforward blasting nastiness for our **No.2** metal meanie, with Doctor Who's old adversaries the **Daleks**. With their battle cry of 'Exterminate!' they've been putting fear into humans of all ages since 1963. They are the only monsters the Doctor ever encountered who got the omniscient Time Lords so rattled that they foresaw a time when the pepper-pots would have annihilated all other races – and so sent our hero back in time in a naughty rule-breaking mission to prevent, or at least hinder, the Daleks' creation.

And our **No.1**? It has to be an automaton so focused, so implacable, so unstoppable that it looks as if there's going to be no way to defeat it. One which just keeps coming, no matter what you throw at it. One which never weakens, never tires, and which just focuses on its mission to kill. The ultimate lean, mean, deadly killing machine is of course **The Terminator**. And if the Arnie variety looked a bit clunky by 1991 along came the T-1000 with its smooth liquid morphing into everything from a policeman to a pickaxe – not to mention some very fast running – to terrify us all over again. Don't have nightmares. He'll be back.

Chapter 15

Loving The Robot

The greatest ever Robot Romances and other Emotional Encounters
Robots, so the assumed wisdom goes, are meant to be stoically immune to any emotional trauma, cold creatures of pure logic. Their lack of any fleshy component – unlike their relatives the Cyborgs, whose origins are partly organic – means that it is taken as a given, most of the time, that they don't feel and suffer and react emotionally like human beings. If no heart beats beneath the steel plates, then there can be no love. It's a romantic reading, and one which doesn't necessarily take account of the way emotional intelligence can develop. Surely, if robots can develop to the point where they mimic human appearance and behaviour – and bearing in mind some recent Japanese innovations like Project Aiko, it is clear that we are heading rapidly towards that point – then surely the nanobots and microprocessors are capable of firing up and activating something very akin to a biochemical reaction? That old-fashioned coup de foudre – that old devil, no less, called love? Let's take a look at some of the instances of robots behaving almost like human beings – and, in the process, perhaps very nearly having their metallic hearts broken.

NATHANAEL and OLYMPIA (The Sandman, 1817 story)
The obsession of artistic Nathanael for a robotic facsimile is grotesquely comic – taken at face value, it's hard to see how a man could love a stiffly-moving, squawking parody of the human form. But the robo-love is here being used to satirical effect; Hoffmann is inviting us to compare Nathanael's idolisation of his beloved Clara, and by extension the idolisation of Woman by Man. Can love be a depersonalising emotion, one which makes human beings sometimes fall in love with the idea of love, almost to the point where the love-object itself is no longer someone with feelings that matter? And therefore may as well, to all intents and purposes, be a robot?

THE TIN MAN'S HEART (The Wizard of Oz, 1939 film)
Only in the books is the Tin Man a former human, one who rather ghoulishly replaced living organs with metal. In the film, he is a rather more audience-friendly (and child-friendly) figure – a metal man who wants a heart, because the tinsmith

'forgot' to give him one. The implications of this are rather interesting. The Tin Man is technically an automaton of some kind, and yet it is presumed that the simple insertion of a heart – one which is, apparently within the wizardly remit, at least as far as the Tin Man himself believes – will enable him to experience greater humanity. It's perhaps just as well that, in the end, the climax of the film rather dodges the question by making the bestowal of a heart rather more of a symbolic gesture.

CORRY and ALICIA (The Twilight Zone: The Lonely, 1959)
Well, this one doesn't end well at all. Criminal James Corry, exiled to a barren world, gets the lovely robot Alicia as a companion, falls in love with her, and is then told by the chap who comes to pick him up after his pardon that (with an echo of classic SF story 'The Cold Equations') she would exceed the weight limit. Luckily, Alicia is only a robot and doesn't take offence at this. Unluckily, she's then summarily shot in the face. The consoling words that 'all you're leaving behind is loneliness' seem a little hollow… Yes, one can look at the issue flippantly, but it raises the question of whether artificial intelligence could ever get to the point where it could experience something like real 'love.'

C-3PO and R2-D2 (Star Wars and sequels, 1977 onwards)
The first characters we see in the film – and among its most likeable. These two can hardly be dismissed as simply walking lash-ups of microprocessors, as it's from the start they are supposed to be characters, and a duo with a relationship of sorts too. While one would hesitate to go as far as to apply the fashionable 21st-century tag of 'Bromance', there's an obvious affection between them underneath the bickering.

THE REPLICANTS (Blade Runner, 1982 film)
The Replicants in Blade Runner are looked upon as one of cinema's most exciting robot innovations, and their complexities are covered in a separate entry here. Even three decades on, the ramifications of the Replicant experience are still hotly debated by fans – not to mention the controversial question of the status of hero/narrator Deckard himself, whose experience with replicant Rachael is played for all-out misty-eyed film noir romance. What's the meaning of that unicorn, the creature of his dreams which appears as an origami figure at the end of the Director's Cut? What can't be denied is that Roy Batty's 'death' scene, as played by Rutger Hauer, is hugely moving. If he truly remembers seeing all the things he claims to have seen – and knows that they are now gone 'like tears in rain' – then who is to say that they are any less 'real'?

LISA (Weird Science, 1985 film)
It's the dream of all nerdy teenage boys – using the computer to create a perfect girlfriend in their bedroom. And this was two decades before Facebook. John

Hughes's teen sci-fi comedy presents us with the ultimate in 'male gaze' love-robot, the virtual perfect woman, partly inspired by a Barbie doll (whose measurements, one must point out, could never exist in real life or she'd always be falling over). And – the ultimate in wish-fulfilment – she will do anything they want. Once again, the robot form of romance is falling rather short of the real thing in its attempts to be an ideal version of it...

DATA's EMOTION CHIP (Star Trek: The Next Generation, 1987 onwards)

It's appropriate that the Utopian, techno-fetishizing Star Trek universe should have as one of its most popular characters the innocent, almost little-boy-lost Data, a robot without affection but also without malice, driven by logic and finding his way in the human world. He's the Star Trek reboot's answer to the emotionless Spock from the 1960s series (who, thanks to his partly human heritage, was occasionally capable of flashing a smile, making an acerbic quip and even – on one memorable occasion – doing what appeared to be falling in love). And it's also typical Star Trek to feature a creation like an 'emotion chip' – something which can simply be plugged into a robot to allow it to experience the complexities of the human heart. But, paradoxically, Data does display behaviour which one could call 'emotional' attachment – he's loyal to his colleagues and to the Federation, for a start, and his general persona is vulnerable, naïve and likeable rather than cold and ruthless. Data's attempts to understand what makes humans tick make for some of The Next Generation's most enjoyable moments. He's seen to show anger, and laughter, and this fits in with creator Gene Rodenberry's stated intention to make Data more 'human' as the series went on. It's not until the film franchise, though, and the original series/Next Generation crossover film Generations, that Data finally gets to have the emotion chip properly installed and experience almost fully what it means to be human.

HUMANS and CYLONS (The re-imagining of TV series Battlestar Galactica, 2004-2009)

OK, so it's hard to imagine any human wanting to get it on with the Cylons from the original 70s show – those clunking, shiny, robot knights with the single red eye uttering their catchphrase 'By your command' in a Metal Mickey voice. But the new 'skinjob' Cylons (a description which references the Blade Runner Replicants) are a different proposition altogether. They're now based on bio-engineering, creatures of synthetic biology. And, just like the Replicants again, some of them are 'sleepers' – minds packed with false memories so that they truly believe themselves to be human...

WALL-E and EVE (2008)
Perhaps cinema's cutest robot-flirtation. It's fun to see the relationship between these two growing, and then 'aaaah' as they dance and hug and 'kiss' – a spark bouncing from Eve's head to Wall-E's – in childlike innocence. Perhaps these robots embody something the human race has lost – emotion unencumbered by cynicism or a sense of discord or betrayal?

LE TRUNG and AIKO (2009)
An odd pairing from real life. Project Aiko, the convincing replica of a woman dubbed a 'fembot' by tabloid reporters ever more anxious to see science-fiction become science-fact in their own lifetime, is the creation of Canadian genius Le Trung. In 2009, he too Aiko home to meet his parents, describing her as 'just like any other girl.' (Yes, except presumably you aren't going to be taking her swimming.) His £30,000 construction apparently helps him choose his food and clothes. Still, who are we 21st-century humans to mock such developments? Maybe Le Trung will have the last laugh. This could be the future of the human race. There may come a time when we exist side-by-side, looking from one to the other, unable to tell, now, which is human and which is robot.

Chapter 16

The Perfect Robot

- First of all, a robot needs tremendous **intelligence and memory power.** We now live in an age which even the tech-obsessed 1980s couldn't imagine, where the entire contents of a bookshelf can be stored on something the size of a thumbnail – the 'memory crystals' used in 90s SF show *Babylon 5* were this size, and they already seem archaic after a couple of decades.
- Whether it's K-9 offering data retrieval or Artoo synching up with the Death Star, the perfect robot will have the capacity to **retrieve and analyse information** in the click of a finger. After all, if your robot isn't brighter or faster than you, then what's the point in having him?
- The perfect robot always has **tremendous strength**, or other superhuman abilities. Humanity hates boring or unpleasant jobs, and so it makes sense that the robots we create, both in fact and in fiction, are capable of feats greater than the human body can manage. Whether it is Gort, impressing with his mere presence and a blast of light from his helmet, or the Terminator surviving the explosion of a petrol-tanker, the Robot is the mechanical embodiment of the 'superman' – like us, but more powerful.
- Alongside this, a robot needs **speed**, like the unstinting running of the T-1000, the shimmeringly fast Sentinels of the Matrix films. The lumbering, chunky automata of earlier decades, like the laughable Annihilants, would not pass muster now, not even as some kind of retro tribute.
- Ideally, the robot will **do no harm.** Asimov's principles have been broken many times in fiction, though, and the danger in making a robot that can think for itself is that we risk the creation of an automaton with something close to free will, which is truly hostile to humanity – whether in self-defence or self-preservation, or with more nefarious intent. Often, in robot fiction, only the robot's creator/programmer is capable of setting it on the right path again.
- A robot companion needs to be **loyal**. Working alongside humanity, there should be no question that it will defect to our enemies. It will work, think and struggle at our side no matter how tough things get. A perfect robot will never lie to you or try to double-cross you, and it will work unceasingly without demanding breaks or going on strike.

- And finally, appropriately for this appearance-obsessed decade, we need the perfect robot to have **aesthetic appeal**. The robots we remember are not just the efficient ones, but those which have a haunting beauty, like the unmistakable Robby, the golden Maria from *Metropolis*, the gleaming C-3PO, the fluid T-1000 Terminator and the almost-human Replicants. Current science is obsessed with producing the perfect anthropomorphic robot, coming close with developments like the Japanese robot pop star HRP-4C. Is the ideal robot perhaps a **shape-shifter**, able to take any form chosen and pre-programmed by its human masters? (Although this could cause no end of embarrassment at cocktail parties. If you thought turning up in the same dress as the hostess was bad enough, what if two of you turn up with the same companion?)
- Yes, it is surely inevitable that, as technology progresses, we will move ever closer to robots which are **indistinguishable from human beings**, and whose microprocessors, receptors and circuits mimic near-perfectly the workings of the synapses and neurons in the brain. Where, then, will we be left? Perhaps we will be in need of a whole new set of laws and rules to govern the rights of our robot brethren… Until that day comes, a fond farewell for now from robot-world…